Nuclear Power in the Developing World

KING ALFRED'S COLLEGE
WINCHESTER

To be returned on or before the day marked
below :—

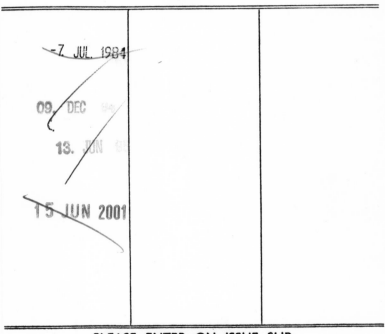

-7 JUL. 1984

09. DEC

13. JUN

15 JUN 2001

PLEASE ENTER ON ISSUE SLIP:

AUTHOR   PONEMAN

TITLE   Nuclear power in the developing world

ACCESSION No.   1008

Nuclear Power in the Developing World

DANIEL PONEMAN

Research Fellow
Center for Science & International Affairs, Harvard University

London
GEORGE ALLEN & UNWIN
Boston   Sydney

George Allen & Unwin (Publishers) Ltd,
40, Museum Street, London, WC1A 1LU, UK

George Allen & Unwin (Publishers) Ltd,
Park Lane, Hemel Hempstead, Herts, HP2 4TE, UK

Allen & Unwin Inc.,
9 Winchester Terrace, Winchester, Mass 01890, USA

George Allen & Unwin Australia Pty Ltd,
8 Napier Street, North Sydney, NSW 2060, Australia

First published in 1982

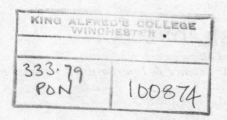
British Library Cataloguing in Publication Data

Poneman, Daniel
  Nuclear power in the developing world
1. Atomic power — Underdeveloped areas
I. Title
621.48′ 091724   TK9145
ISBN 0-04-338100-6

Library of Congress Cataloging in Publication Data

Poneman, Daniel
  Nuclear power in the developing world
Bibliography: p.
1. Underdeveloped areas — Atomic power industry
I. Title
HD9698.A2P58 1982   333.79′ 24′ 091724   82-11367
ISBN 0-04-338100-6

Set in 10 on 11 point Times by Typesetters (Birmingham) Ltd
and printed in Great Britain
by Billing & Sons Ltd, London and Worcester

Contents

To My Parents and Grandparents

Foreword

Who, other than the handful of companies seeking to sell nuclear power plants and the billions of people living in the developing world, cares about the subject matter of this book? Who need to improve their understanding of the ways in which developing nations formulate and implement their nuclear policies? Everyone has an important stake in these matters, especially those who care about how governments make decisions, about meeting the world's energy needs in the decades ahead, about the environmental and public health implications of increased (or decreased) use of nuclear power, and about the proliferation of nuclear weapons. Each country's nuclear policy affects every country's future.

National choices of nuclear policies involve consideration of diverse and often conflicting objectives. Credible arguments for increased use of nuclear power can be based upon economics, public health and safety, environmental protection, foreign policy, and national security. The same factors can be employed to construct comparably credible arguments in opposition to nuclear power. For example, in some situations increased use of nuclear power can reduce the need for imported oil – a clear economic advantage in terms of balance of international trade; but the cost of nuclear-generated electricity may be higher than that produced by the burning of oil – an equally clear economic disadvantage. Under normal operation, nuclear power plants are likely to have less adverse effects upon public health and safety and the environment than coal-burning plants; but, if severe nuclear accidents turn out to be more likely than anticipated by 'the experts', the consequences of following the nuclear path could be catastrophic. Reduced dependence on imported fossil fuels could provide greater freedom of choice in foreign policy, especially in relations with oil-exporting countries, but increased dependence on foreign nuclear technology and imported reactor fuel can only restrict foreign policy options. Acquisition of nuclear materials and facilities that could be useful in the production of nuclear weapons might ultimately enhance national security, but it might instead stimulate a preemptive attack by a suspicious adversary.

In this study, Daniel Poneman examines objectively and comprehensively the panoply of goals and influences bearing upon nuclear policymaking. His description of the requirements for a nuclear power program and of the current status of the global nuclear enterprise is presented in language fully understandable by the nontechnical reader; his case studies (of the nuclear programs in three developing

countries: Argentina, Iran, and Indonesia) are particularly well
selected and presented; his analysis is original and insightful; and his
conclusions and recommendations deserve the attention of those who
care about our nuclear future.

Daniel Poneman is my student, my colleague, my friend, and my
teacher. This is his first book. I am confident that readers will join me
in hoping that there will be many more.

ALBERT CARNESALE
Cambridge, Massachusetts
January 1982

Acknowledgements

This book could not have been completed without the help of many teachers, colleagues, and friends. Special thanks go to those who commented on the manuscript, especially Albert Carnesale, Alan Albert, Philip Bobbitt, Lynn Davis, Timothy George, Robert Litwak, Robert Shishko, and Gregory Treverton. The impetus behind the project was supplied by Allan Pulsipher of the Ford Foundation and Nicholas Brealey of George Allen & Unwin. The book grew out of an Oxford thesis, which benefited from the trenchant insights of my supervisor, Hedley Bull. I would also like to thank past teachers Paul Doty, Joseph Nye, Michael Nacht, David Goldey, and Robert Gallucci. Senator John Glenn initiated my interest in the nuclear issue, six years ago. Ever since, I have been extremely lucky to have enjoyed the counsel, humor, and constant encouragement of Albert Carnesale.

I am deeply grateful to the Ford Foundation, whose generosity enabled me to devote a year to the project and to conduct research in Europe, North and South America, and Asia. Without Jerry Scott in Buenos Aires, Terance Bigalke in Jakarta, and Solita del Castillo in Manila, I would have been lost. Christoph Bertram kindly offered ideal working conditions at the International Institute for Strategic Studies, London, during the 1980–81 academic year. The IISS staff offered unstinting assistance. The Cyril Foster Fund and Lincoln College, Oxford University, also provided research support. Respect for their confidentiality bars individual acknowledgment of the many interviewees who graciously endured lengthy questionings and provided invaluable information. Thanks are also due to the Royal Institute of International Affairs for permission to borrow from my article in the Autumn 1981 issue of *International Affairs*. Rachel Smith and Eileen Webb typed the manuscript expertly and cheerfully. Rowena Friedman meticulously prepared the copy for typesetting. Yeo Puay Huei helped with the Indonesian translations. Finally, I would like to express my deepest gratitude to family and friends. I am, of course, responsible for all of the book's shortcomings.

D.P.
Cambridge, Massachussetts
November 1981

List of Tables

List of Figures

List of Abbreviations

AEC	–	Atomic Energy Commission
AECL	–	Atomic Energy of Canada Limited
AEOI	–	Atomic Energy Organization of Iran
BARC	–	Babha Atomic Research Centre
BAPPENAS	–	Indonesian state planning agency
BATAN	–	Indonesian National Atomic Power Agency
CANDU	–	Canadian deuterium uranium reactor
CAS	–	Committee for Assurance of Supply
CEA	–	French Commissariat for Atomic Energy
CIRUS	–	Canadian–Indian–US research reactor
CMEA	–	Council for Mutual Economic Assistance (Comecon)
CNEA	–	Argentine National Atomic Energy Commission
CNEN	–	Brazilian National Nuclear Energy Commission
COFACE	–	French Company for the Insurance of External Commerce
CONAES	–	Committee on Nuclear and Alternative Energy Systems
CTBT	–	Comprehensive Test Ban Treaty
DAE	–	Indian Department of Atomic Energy
DM	–	Deutschmark
EdF	–	Electricité de France
ENACE	–	Argentine Nuclear Enterprise for Electrical Stations
EPEC	–	Argentine National Executive Power
Euratom	–	European Atomic Energy Community
Eurodif	–	French-led uranium enrichment consortium
FBIS	–	Foreign Broadcast Information Service
FF	–	French franc
GDP/GNP	–	gross domestic product/gross national product
GW	–	gigawatt (= 1,000 MW)
HWR	–	heavy water reactor
IAEA	–	International Atomic Energy Agency
IBRD	–	International Bank for Reconstruction and Development (World Bank)
INFCE	–	International Nuclear Fuel Cycle Evaluation
KANUPP	–	Karachi Nuclear Power Plant
KECO	–	Korean Electric Company
KFW	–	West German Credit Bank for Reconstruction
kWh	–	kilowatt (hours)
KWU	–	Kraftwerk Union
LTBT	–	Limited Test Ban Treaty
MIGAS	–	Indonesian Oil and Gas Directorate

MW (e or t)	–	Megawatt (electrical or thermal)
MAPS	–	Madras Atomic Power Station
NAPS	–	Narora Atomic Power Station
Nira	–	Italian Company for Advanced Nuclear Reactors
NNPA	–	Nuclear Nonproliferation Act
NNWS	–	nonnuclear weapon state
NPT	–	Treaty on the Non-Proliferation of Nuclear Weapons
NRC	–	Nuclear Regulatory Commission
NSG	–	Nuclear Suppliers Group (London Club)
NSSS	–	nuclear steam supply system
Nuclebras	–	Brazilian state nuclear company
NWFZ	–	nuclear weapon free zone
NWS	–	nuclear weapon state
OPEC	–	Organization for Petroleum Exporting Countries
PBO	–	Iranian Planning and Budget Organization
PDV	–	present discounted value
Pertamina	–	Indonesian state oil company
PLN	–	Indonesian state electrical utility
PN Batubara	–	Indonesian state coal company
PNE	–	peaceful nuclear explosion
PNPP	–	Philippine Nuclear Power Plant
PWR	–	pressurized water reactor
RAPS	–	Rajasthan Atomic Power Station
REPELITA	–	Indonesian five-year development plan
RIIA	–	Royal Institute of International Affairs
SWU	–	separative work unit
Taipower	–	Taiwan Power Company
TAPS	–	Tarapur Atomic Power Station
Tavanir	–	Transmission and Distribution Company of Iran
T&D	–	transmission and distribution
TMI	–	Three Mile Island
Triga	–	an American-made research reactor
TW	–	terawatt ($= 1,000$ GW)
UNDP	–	United Nations Development Program
UNIDO	–	United Nations Industrial Development Organization
Uramex	–	Mexican state uranium company
Urenco	–	Anglo-Dutch-German uranium enrichment consortium
WASP	–	Wien Automatic System Planning Package

Nuclear Power in the Developing World

Nuclear Power in the Developing World

DANIEL PONEMAN

Research Fellow
Center for Science & International Affairs, Harvard University

London
GEORGE ALLEN & UNWIN
Boston   Sydney

George Allen & Unwin (Publishers) Ltd,
40, Museum Street, London, WC1A 1LU, UK

George Allen & Unwin (Publishers) Ltd,
Park Lane, Hemel Hempstead, Herts, HP2 4TE, UK

Allen & Unwin Inc.,
9 Winchester Terrace, Winchester, Mass 01890, USA

George Allen & Unwin Australia Pty Ltd,
8 Napier Street, North Sydney, NSW 2060, Australia

First published in 1982

British Library Cataloguing in Publication Data

Poneman, Daniel
  Nuclear power in the developing world
1. Atomic power — Underdeveloped areas
I. Title
621.48′ 091724   TK9145
ISBN 0-04-338100-6

Library of Congress Cataloging in Publication Data

Poneman, Daniel
  Nuclear power in the developing world
Bibliography: p.
1. Underdeveloped areas — Atomic power industry
I. Title
HD9698.A2P58 1982   333.79′ 24′ 091724   82-11367
ISBN 0-04-338100-6

Set in 10 on 11 point Times by Typesetters (Birmingham) Ltd
and printed in Great Britain
by Billing & Sons Ltd, London and Worcester

Contents

To My Parents and Grandparents

Foreword

Who, other than the handful of companies seeking to sell nuclear power plants and the billions of people living in the developing world, cares about the subject matter of this book? Who need to improve their understanding of the ways in which developing nations formulate and implement their nuclear policies? Everyone has an important stake in these matters, especially those who care about how governments make decisions, about meeting the world's energy needs in the decades ahead, about the environmental and public health implications of increased (or decreased) use of nuclear power, and about the proliferation of nuclear weapons. Each country's nuclear policy affects every country's future.

National choices of nuclear policies involve consideration of diverse and often conflicting objectives. Credible arguments for increased use of nuclear power can be based upon economics, public health and safety, environmental protection, foreign policy, and national security. The same factors can be employed to construct comparably credible arguments in opposition to nuclear power. For example, in some situations increased use of nuclear power can reduce the need for imported oil – a clear economic advantage in terms of balance of international trade; but the cost of nuclear-generated electricity may be higher than that produced by the burning of oil – an equally clear economic disadvantage. Under normal operation, nuclear power plants are likely to have less adverse effects upon public health and safety and the environment than coal-burning plants; but, if severe nuclear accidents turn out to be more likely than anticipated by 'the experts', the consequences of following the nuclear path could be catastrophic. Reduced dependence on imported fossil fuels could provide greater freedom of choice in foreign policy, especially in relations with oil-exporting countries, but increased dependence on foreign nuclear technology and imported reactor fuel can only restrict foreign policy options. Acquisition of nuclear materials and facilities that could be useful in the production of nuclear weapons might ultimately enhance national security, but it might instead stimulate a preemptive attack by a suspicious adversary.

In this study, Daniel Poneman examines objectively and comprehensively the panoply of goals and influences bearing upon nuclear policymaking. His description of the requirements for a nuclear power program and of the current status of the global nuclear enterprise is presented in language fully understandable by the nontechnical reader; his case studies (of the nuclear programs in three developing countries: Argentina, Iran, and Indonesia) are particularly well selected and presented; his analysis is original and insightful; and his conclusions and recommendations deserve the attention of those who care about our nuclear future.

Daniel Poneman is my student, my colleague, my friend, and my teacher. This is his first book. I am confident that readers will join me in hoping that there will be many more.

ALBERT CARNESALE
Cambridge, Massachusetts
January 1982

Acknowledgements

This book could not have been completed without the help of many teachers, colleagues, and friends. Special thanks go to those who commented on the manuscript, especially Albert Carnesale, Alan Albert, Philip Bobbitt, Lynn Davis, Timothy George, Robert Litwak, Robert Shishko, and Gregory Treverton. The impetus behind the project was supplied by Allan Pulsipher of the Ford Foundation and Nicholas Brealey of George Allen & Unwin. The book grew out of an Oxford thesis, which benefited from the trenchant insights of my supervisor, Hedley Bull. I would also like to thank past teachers Paul Doty, Joseph Nye, Michael Nacht, David Goldey, and Robert Gallucci. Senator John Glenn initiated my interest in the nuclear issue, six years ago. Ever since, I have been extremely lucky to have enjoyed the counsel, humor, and constant encouragement of Albert Carnesale.

I am deeply grateful to the Ford Foundation, whose generosity enabled me to devote a year to the project and to conduct research in Europe, North and South America, and Asia. Without Jerry Scott in Buenos Aires, Terance Bigalke in Jakarta, and Solita del Castillo in Manila, I would have been lost. Christoph Bertram kindly offered ideal working conditions at the International Institute for Strategic Studies, London, during the 1980–81 academic year. The IISS staff offered unstinting assistance. The Cyril Foster Fund and Lincoln College, Oxford University, also provided research support. Respect for their confidentiality bars individual acknowledgment of the many interviewees who graciously endured lengthy questionings and provided invaluable information. Thanks are also due to the Royal Institute of International Affairs for permission to borrow from my article in the Autumn 1981 issue of *International Affairs*. Rachel Smith and Eileen Webb typed the manuscript expertly and cheerfully. Rowena Friedman meticulously prepared the copy for typesetting. Yeo Puay Huei helped with the Indonesian translations. Finally, I would like to express my deepest gratitude to family and friends. I am, of course, responsible for all of the book's shortcomings.

D.P.
Cambridge, Massachussetts
November 1981

List of Tables

List of Figures

List of Abbreviations

AEC   – Atomic Energy Commission
AECL   – Atomic Energy of Canada Limited
AEOI   – Atomic Energy Organization of Iran
BARC   – Babha Atomic Research Centre
BAPPENAS   – Indonesian state planning agency
BATAN   – Indonesian National Atomic Power Agency
CANDU   – Canadian deuterium uranium reactor
CAS   – Committee for Assurance of Supply
CEA   – French Commissariat for Atomic Energy
CIRUS   – Canadian–Indian–US research reactor
CMEA   – Council for Mutual Economic Assistance (Comecon)
CNEA   – Argentine National Atomic Energy Commission
CNEN   – Brazilian National Nuclear Energy Commission
COFACE   – French Company for the Insurance of External Commerce
CONAES   – Committee on Nuclear and Alternative Energy Systems
CTBT   – Comprehensive Test Ban Treaty
DAE   – Indian Department of Atomic Energy
DM   – Deutschmark
EdF   – Electricité de France
ENACE   – Argentine Nuclear Enterprise for Electrical Stations
EPEC   – Argentine National Executive Power
Euratom   – European Atomic Energy Community
Eurodif   – French-led uranium enrichment consortium
FBIS   – Foreign Broadcast Information Service
FF   – French franc
GDP/GNP   – gross domestic product/gross national product
GW   – gigawatt (= 1,000 MW)
HWR   – heavy water reactor
IAEA   – International Atomic Energy Agency
IBRD   – International Bank for Reconstruction and Development (World Bank)
INFCE   – International Nuclear Fuel Cycle Evaluation
KANUPP   – Karachi Nuclear Power Plant
KECO   – Korean Electric Company
KFW   – West German Credit Bank for Reconstruction
kWh   – kilowatt (hours)
KWU   – Kraftwerk Union
LTBT   – Limited Test Ban Treaty
MIGAS   – Indonesian Oil and Gas Directorate

MW (e or t)	–	Megawatt (electrical or thermal)
MAPS	–	Madras Atomic Power Station
NAPS	–	Narora Atomic Power Station
Nira	–	Italian Company for Advanced Nuclear Reactors
NNPA	–	Nuclear Nonproliferation Act
NNWS	–	nonnuclear weapon state
NPT	–	Treaty on the Non-Proliferation of Nuclear Weapons
NRC	–	Nuclear Regulatory Commission
NSG	–	Nuclear Suppliers Group (London Club)
NSSS	–	nuclear steam supply system
Nuclebras	–	Brazilian state nuclear company
NWFZ	–	nuclear weapon free zone
NWS	–	nuclear weapon state
OPEC	–	Organization for Petroleum Exporting Countries
PBO	–	Iranian Planning and Budget Organization
PDV	–	present discounted value
Pertamina	–	Indonesian state oil company
PLN	–	Indonesian state electrical utility
PN Batubara	–	Indonesian state coal company
PNE	–	peaceful nuclear explosion
PNPP	–	Philippine Nuclear Power Plant
PWR	–	pressurized water reactor
RAPS	–	Rajasthan Atomic Power Station
REPELITA	–	Indonesian five-year development plan
RIIA	–	Royal Institute of International Affairs
SWU	–	separative work unit
Taipower	–	Taiwan Power Company
TAPS	–	Tarapur Atomic Power Station
Tavanir	–	Transmission and Distribution Company of Iran
T&D	–	transmission and distribution
TMI	–	Three Mile Island
Triga	–	an American-made research reactor
TW	–	terawatt (= 1,000 GW)
UNDP	–	United Nations Development Program
UNIDO	–	United Nations Industrial Development Organization
Uramex	–	Mexican state uranium company
Urenco	–	Anglo-Dutch-German uranium enrichment consortium
WASP	–	Wien Automatic System Planning Package

Part One: Issues

1 Introduction

Developing country governments may want to acquire nuclear technology for several reasons. They may wish to use nuclear power stations to increase their electrical generating capacity, or to develop the capacity to build nuclear weapons, or simply to create the option to pursue either energy or military routes as future policy requires. It is also possible that nuclear policies are driven less by rational choice than by domestic political considerations. Pressure from foreign governments and corporations may encourage nuclear programs where otherwise there would be little interest in fission. At the same time, any of these considerations – energy or military objectives, domestic politics, and foreign influences – can discourage interest in nuclear power in any country.

The purpose of this book is to explore why developing country governments choose the nuclear policies they do. This issue must be squarely faced because several developments in the last few years have occasioned grave concern over the possibility that developing country nuclear energy programs will evolve into military programs. First, on May 18, 1974, the Indian government detonated a 'peaceful' nuclear explosive device (PNE) at the bottom of a 107-meter shaft at Pokharan in the Rajasthan Desert. The plutonium core of the device was produced in an ostensibly civil nuclear program. All other members of the nuclear club – the United States, the United Kingdom, the Soviet Union, France, and China – entered that elite through explicitly military efforts. Fears multiplied that other developing countries would follow the Indian cue in abusing nuclear technology obtained through foreign assistance.

Second, increased technical capabilities in some countries generated concern that soon their scientists would be able to process atomic weapons-grade uranium and plutonium. The apparent willingness of some in the advanced nations to sell these nuclear fuel processing technologies to the developing world aggravated these concerns. In 1975 and 1976, West Germany agreed to sell Brazil eight reactors *and* the facilities required for production of enriched uranium and plutonium, and France agreed to sell plutonium reprocessing plants to Pakistan and South Korea. Belief spread that the Atoms for Peace premise, that technology transfers would be safe so long as safeguards were applied, had been ill-conceived.

Third, the June 7, 1981 attack on Iraqi nuclear facilities reflected the creeping apprehension that even a government with impeccable nonproliferation credentials could develop nuclear weapons. Iraq is a member of the Treaty on the Non-Proliferation of Nuclear Weapons (NPT) and has accepted International Atomic Energy Agency (IAEA) safeguards over all of its nuclear activities. For all that, many believed along with the Israelis that President Saddam Hussein intended to build a military nuclear program. Iraqi insistence that France supply it with weapons-grade, 90 percent enriched uranium supported this view. The Israeli attack, however, highlighted the prospect not only of governments openly accepting peaceful use constraints while secretly harboring military aims, but also of incredulous third parties taking matters into their own hands, initiating conflicts that could draw in the major powers.

Growing concern over developing country nuclear intentions led the suppliers of nuclear technology to begin restricting exports, lest assistance be diverted covertly to explosive use. Following a Kissinger initiative, a Nuclear Suppliers Group (NSG) began meeting in London in 1975 in order to agree on a set of guidelines that would restrict the export of certain sensitive items and require more extensive use of international safeguards. The United States, followed by France and West Germany, forswore the export of plutonium reprocessing technology. The Canadians used the threat of suspended technical assistance and natural uranium supplies to impose stricter nonproliferation controls on Argentina, India, and Pakistan. Supplier policy decisively affects the developing countries because of their extensive dependence upon outside help.

Increased concern over Third World nuclear policies both reflected and contributed to a loss of confidence in the IAEA safeguards system. Behind the creation of the NSG lay the notion that IAEA safeguards could not adequately guarantee the peaceful use of nuclear technology. The utility of any international system, which survives only by grace of the confidence participants place in it, dwindles as factions develop within it and seek to go beyond its mandate. What one group sees as supplementing the system, another sees as circumventing it. The nuclear suppliers (and particularly the United States) faced a dilemma: having promoted the creation of an organization which had gained the acceptance of 110 member states, how could they significantly increase its restrictions on nuclear exports without either offering a corresponding increase in promotional nuclear exports, or angering the vast majority of IAEA members which were nuclear technology importers? Conceding that the agency alone could not enforce the increased restrictions which many felt were necessary after 1974, it was extremely difficult for the supplier governments to go further without undermining the consensus which had already been achieved. The deep differences of opinion dividing the European from the American suppliers only aggravated the problem.

Building any consensus on a sensible solution depends upon agreeing on the nature of the problem: why do various developing country governments really want nuclear power? To assist in this effort, this book will try to answer the following central questions. First, what traits of nuclear power are most important in the developing world? Second, what sorts of nuclear policies do different developing countries adopt? Third, what do these policies tell us about the driving forces behind them?

Each of the three sections of this book is addressed to one of these questions. Part One includes a brief sketch of the key features of nuclear power as they relate to its perceived utility (Chapter 2). Part Two moves from generic assessment to concrete examples of how various developing countries approach the nuclear option (Chapter 3). A glaring omission in some studies is the failure to include governments which have *not* yet chosen to build nuclear power stations. After all, to reject or defer the nuclear option in favor of others is just as much a decision as to adopt it. Of the developing countries which have pursued nuclear power programs a broad distinction is drawn between two approaches. In one, called 'independent', governments stress the contribution made by nuclear power to economic and political independence, seek to escape reliance on outside suppliers by achieving self-sufficiency in the nuclear fuel cycle, and sometimes adopt strident or at least stubborn attitudes in nonproliferation discussions. In the second, called 'dependent', governments try to bring as much nuclear power on line as quickly as possible with extensive outside help, and appease the nuclear suppliers by accepting international safeguards over all nuclear activities and *not* pursuing fuel cycle independence.

In order to avoid overgeneralization, three case studies are presented, revealing firsthand the anatomy of nuclear policy development in each approach. Argentina (Chapter 4) represents the 'independent' style of development; Iran (Chapter 5), the 'dependent'; and Indonesia (Chapter 6), the 'nonnuclear' approach. (Readers may omit these three chapters without losing the thread of the argument.) The sample countries are sufficiently similar to justify comparison. All are rich in fossil fuel resources; two are also well endowed with hydropower potential. Despite disparities in income and industrial capabilities, all three have suffered from shortages and outflows of talent and, usually, from economic distress. All three programs received impetus from US

Atoms for Peace grants. Each has been administered by an atomic energy commission. All three governments have resorted to national energy plans and, in each, nuclear power was advocated vigorously. There the similarities end. The three followed widely divergent nuclear policies. The Argentine nuclear program commenced around 1950. It included the indigenous construction of several research reactors and led to the operation of the first power reactor in Latin America. The Iranian program lay dormant for over fifteen years before the Shah decreed that it quickly become one of the world's largest. In Indonesia, early interest in nuclear energy has yet to result in the implementation of a serious program.

These three were selected not as typical, but as illustrative; the existing literature and available data guided the choice. So much has already been written about India and Brazil that they were excluded from special consideration, despite their prominence in the developing world nuclear scene. Indonesia was included precisely because of the lack of attention accorded its nuclear program. The moribund Iranian program was more accessible than many others, having been stripped of the vested interests which so often hide the sources of decision.

Part Three explores four possible explanations for nuclear policies, drawing upon several examples but especially upon the three featured in Part Two. Nuclear power may be desired by virtue of its security benefits (Chapter 7). Governments in precarious positions may seek to cultivate fission technology in the hope of shoring up public support. Because fission technology may be applied toward military ends, governments may wish to use its acquisition to provide leverage in bargaining with countries, near and far, that oppose the spread of nuclear weapons. Nuclear weapons themselves provide the most menacing political incentive for the acquisition of civil nuclear technology.

Economics is nearly always invoked as the primary justification for nuclear power programs (Chapter 8). Some view fission as a well-tried means to expand electricity supplies, desirable for improving both living standards and industrial production capabilities. Proponents commend it as competitive with energy alternatives in price and environmental soundness. Its attraction is enhanced where it can substitute for the use of oil in electricity generation. In the long run, the technical expertise gained is hoped to benefit other industrial activities to which it may be applied.

The chapters on security and economic interests and nuclear development concentrate on objectives to explain policies. In some cases, though, the decision to pursue a nuclear power program may not be reached through a deliberate calculation of costs, benefits, means, and ends. Instead, process may subordinate substance.

Nuclear plans may represent the outcome of bureaucratic battles within government or partisan lobbying from without. Of course, domestic political considerations shape all important government decisions; the issue to be addressed here is whether such considerations produce policies inconsistent with the expressed objectives of a government (Chapter 9). Similarly, one may try to explain policies as the outcomes of foreign influences (Chapter 10). Does pressure from large corporations and their home governments, eager to expand markets in the developing world, succeed in forcing the adoption of nuclear power schemes by Third World leaders who are skeptical of the utility of fission technology for their poverty-stricken economies? Alternatively, does pressure from strengthened environmental concerns abroad prevent these leaders from acquiring a technology which could provide enormous benefit for coming generations?

The conclusion argues that, on balance, objectives prevail over processes in shaping developing country nuclear policies (Chapter 11). The security and economic foundations of a program are mutually dependent. Where potential, long-term economic benefits drive a policy, nuclear planners face the problem of maintaining governmental support in the short term. Surrounding a nuclear power plan with an aura of security or prestige can provide immediate political returns for an undertaking which holds only distant prospects of economic return, thereby dissuading government leaders from abandoning atomic energy. While political concerns are central, economic concerns cannot be ignored. Nuclear power is far too expensive to be justified on vague national security considerations alone. Thus, a plan of long-term economic development through the installation of nuclear power stations remains politically necessary.

Research for this study could not produce incontestable answers. Cabinet minutes and other internal memoranda, which unambiguously could show how policies developed, are and probably will remain inaccessible. Fortunately, a valuable study can be built upon other sources. The histories of the nuclear power policies have been pieced together from official statements, documents, and submissions to international symposia; existing studies and documents from other countries and international organizations; and press reports. Interview, and correspondence with scores of present and former participants in the programs provided invaluable guidance. Their biases could be compensated for but not eliminated, so interview material is used cautiously. Where inferences are inescapable, an effort has been made to avoid compounding misconceptions by building conclusions upon them.

Several terms require definition. In this book, the terms developing world or countries, Third World, and South are used interchangeably to refer to those countries listed as 'developing' by the World Bank, including low- and middle-income and other capital-surplus oil exporting categories. Collective reference to the 'South' does not imply that the governments in it are any more similar than those of the 'North', or that they behave collectively. These terms express neither judgment nor insult. The phrase 'Third World', for instance, does *not* imply inferiority to the 'First' or 'Second'. Offensiveness is a matter of taste and no effort is made here to keep up with fashions. Countries from the Council for Mutual Economic Assistance (Comecon) are also excluded because they are altogether different, hard to penetrate, and less susceptible to the Western and Japanese nuclear suppliers.

A 'nuclear country' is one in which a nuclear power station is under construction or in operation. It may seem unfair to classify as 'nonnuclear' a country like Egypt, with its long operating research reactor and recent conclusion of a contract for two 900 MWe Westinghouse light water reactors. Still, in this area of unfulfilled pledges and pervasive delays, a power program must be seen to be believed. 'Nuclear suppliers' are those countries whose governments and industries transfer nuclear technology and equipment to others, mainly North America, Europe, and Japan, though Argentina and India are beginning to enter this sphere. Exporters of natural uranium who do not engage in other nuclear exports – such as Australia, South Africa, and Niger – are not considered nuclear suppliers. 'Nuclear power' refers to electricity generation for commercial use, whereas 'nuclear energy' and 'nuclear technology' refer more generally to studies, laboratory or prototype scale efforts, medical and agricultural use of radioisotopes, and so on. 'Billion' follows American usage, equaling 1,000 million in British terms. Except where otherwise noted, dollars refer to US currency, unadjusted for subsequent inflation or exchange rate fluctuations.

Finally, several limitations must be noted. The taxonomy described in Part Two is grossly oversimplified. Classification unavoidably caricatures reality, the moreso the fewer the categories. Treating each case as unique, though, so inundates one with data that general trends, if any there be, become obscured in excessive detail. Here a middle course is sought, coarsely dividing the developing world into three 'types' in order to capture essential behavioral differences while leaving enough leeway within each grouping to give due notice to the uniqueness of each country's situation.

This book explores neither the process and causes of nuclear weapons proliferation, nor the overall political systems of the countries discussed. The conclusions pertain exclusively to comparative nuclear power policies, and so can be expected to illuminate only those aspects of the policy process which are related directly to nuclear power. The book makes no judgments about nuclear power itself, from technical, economic, or environmental standpoints. It enquires only into the reasons for, not the wisdom of, different approaches to nuclear development. Similar judgments concerning the wisdom of the exploitation of alternative energy sources or of various energy strategies are beyond the scope of a book of such wide geographic spread. These judgments must all vary by country. Finally, this is not a reference book. Readers interested in deeper treatment of the many fields touched upon should refer to the bibliography. The information and arguments in this book have been selected to explore in brief compass the reasons behind developing country nuclear policies, without trying to be comprehensive.

Despite these limitations, it is hoped that the book has some worth. The existing literature on the nuclear issue in developing countries has certain shortcomings. The understandable concern about future Third World nuclear developments often leads to emphasis on today's salient issues without understanding their historical sources. Great interest surrounds the latest proposals for reprocessing bans and multinational fuel cycles, the Pakistani quest for fissile materials, the source of the atomic explosive-like flash spotted over the South Atlantic on September 22, 1979, the preparation of a new test site in India, the confrontation between Iraq and Israel. Peering too far into the future or leaders' psyches, however, often proves feckless. The classic example is the ever-receding forecast of the first Pakistani nuclear test, always eighteen months or so from 'now'. Without diverting attention from pressing problems, this book seeks to retreat from speculation about what will happen next, addressing instead the motives which lay behind past developing country nuclear policies. This approach may better enable policy-makers to head off problems before they reach the 'crisis management' stage.

This book also differs from previous books dealing with the subject of nuclear power in the Third World, by starting from the premise that nuclear power is the exception there. Many authors, when discussing nuclear power and the developing countries, naturally concentrate on those countries which have nuclear programs. Most developing countries, however, do not, and by distinguishing these more explicitly from those that do, a clearer picture emerges of just how difficult it is to sustain a nuclear power program in the developing world and how powerful a national commitment it requires.

The greatest hope for this book is that it might give some guidance on how to achieve some consistency in nuclear export policies without aggravating relationships either among the suppliers or between North and South. Consistency cannot easily be achieved so long as perceptions remain confused, not only as to the motives behind nuclear power policies, but also as to the causes of nuclear weapons proliferation and the best way to confront them. Dependence of South upon North turns confusion into a festering source of discord. Patently peaceful motivations encourage suppliers to relax export policies and vice versa. This is as it should be. If all political leaders of a given country have long been opposed to the acquisition of nuclear weapons in that country, then to hamstring their civil nuclear plans with harsh antiproliferation restrictions could provoke them unnecessarily, harming both the nonproliferation cause and the diplomatic relations between the governments concerned. Conversely if leaders secretly harbor designs to exploit fission for military purposes, harsh antiproliferation restrictions would be justified; the resentment caused would be unfortunate but necessary. It is hoped that this study will shed some light on how developing country nuclear policies are formulated so as to help devise responses which are effective yet do not appear to be wantonly discriminatory.

2 The Nuclear Option

The most popular use of nuclear energy today is in electricity generation, though the possibility of channeling nuclear process heat directly
to nearby homes and factories is also being explored. Nuclear power
generally provides base load capacity. Base load stations run continuously, to meet the minimum level of demand. But electrical demand
varies hourly, diurnally, and seasonally, so intermediate and peak
load stations are switched on as demand rises (for instance, in the
evening hours, when lights and appliances are in use, or in the
summer, when air conditioners are in use). Older, less reliable and
efficient fossil fuel plants, which are more expensive to run, are
reserved where possible for peak uses. The comparatively low operating costs of hydro and nuclear plants well suit them to base load use.

Today, nuclear power contributes only about 2 percent of primary
energy supply and about 7 percent of electricity generated in the
world.[1] Its developing world contribution is far less. Due to long lead
times, the role played by nuclear power for the balance of this century
will depend upon today's decisions, which in turn are heavily
influenced by the characteristics of fission technology. These can be
grouped under four headings: capital intensity, complexity, fuel
requirements, and special risks.

CAPITAL INTENSITY

Different forms of energy utilize different proportions of labor, land,
and capital. Pre-industrial forms are usually labor-intensive. Conventional power plants are fuel (land)-intensive, increasingly so as fossil
fuel prices climb. Solar collectors, hydro dams, and nuclear reactors
are capital-intensive, requiring heavier investment in construction
than in fuel. In fact, fuel costs are zero for solar and hydropower.
Capital accounts for roughly 70 percent of the cost of electricity
generated in light water reactors, compared to 30 percent for oil and
from 45 to 60 percent for coal (depending on coal quality and
pollution control equipment).[2] A 600 MWe reactor costs three-
quarters, not half, as much as a 1,200 MWe unit.[3] Early hopes for the

commercialization of small (100–200 MWe) reactors led the International Atomic Energy Agency to convene conferences on small and medium power reactors in Vienna (1960) and Oslo (1970). Although smaller reactors were built in a few developing countries (Argentina, India, Pakistan), by the mid-1970s interest had ebbed and reactor manufacturers were offering only 600 MWe or larger units for sale. Technicatome of France developed 125 MWe and 300 MWe models for export, but how many they will be able to sell (if any) remains to be seen.

In an electrical supply network, or 'grid', a single unit should not supply more than 10–15 percent of total capacity, otherwise an unexpected shutdown of that station could trigger a blackout throughout the grid. Consequently, with 600 MWe minimal reactor size, it becomes necessary to have a transmission network connecting around 6,000 MWe generating capacity before nuclear power can be considered viable. Improved management of grid loads and the maintenance of reserve generators, ready to pick up the slack when a nuclear station trips, can reduce but not eliminate the likelihood of blackouts. Also, the larger the unit sizes relative to overall grid capacity, the more reserve generating capacity is required, because one 600 MWe unit is more likely to fail than are three 200 MWe units (reflecting both sheer probability and the greater susceptibility of larger, more complex units to mishap). All reserve generators require maintenance, and 'spinning reserve' (disconnected generators which are kept spinning so as to be able to replace a failed unit immediately) consumes fuel as well, so an increased reserve requirement also raises the per kilowatt cost of delivered electricity. These drawbacks accrue from any large units, nuclear or otherwise, in grids the size of most in the developing world. (See Table 2.1.)

Large power stations also take longer to build. For capital intensive projects, lengthened lead times can significantly increase costs, due both to increases in interest on capital and in the impact of inflation. A long lead time is not in itself harmful; so long as the project proceeds according to schedule, and inflation follows the expected rate, then it may compare favorably with an otherwise identical, shorter lead time project, when both are reduced to their present discounted value (PDV). The problem arises from the unpredictability of longer lead time projects. Inflation may take a heavier toll than expected. If a loan is taken out or a piece of equipment is purchased prior to unexpected delays, then interest on the premature loan and maintenance and warehousing charges for the equipment will also exceed expectations. Added costs from delays arise from penalty clauses in project contracts or, if commencement of reactor operation is postponed, from the need to purchase power from other utilities to fulfill electrical supply commitments. For these reasons, the more accurately a project's completion date can be predicted, the better. In general, this tends to favor shorter lead times and off-the-shelf technologies, for which increased supplier competition also benefits the buyer.

Their size and capital intensity make nuclear projects most attractive in countries with developed industrial and financial sectors. Messrs Reinhold and Schweikert, from the German reactor manufacturer, Kraftwerk Union, described the basic requirements for a nuclear project: a reliable communications system; roads, railways, and shipping routes; transport agencies and harbor facilities, equipped to unload and transport heavy components weighing several hundred tons; residential areas for project staff, complete with markets, schools, hospitals, recreational facilities, sewage treatment, fire brigade, water and power supply; and installations and personnel able to deal with nuclear accidents.[4] The construction of the nuclear station itself requires subcontractors for the supply of raw materials, plates, castings, forgings, pipings, carbon and stainless steel, bearings, seals, flanges, valves, measuring instruments, screws, and bolts. Training centers for skilled workers and plant operators are needed, as are laboratories and construction companies. For a large pressurized water reactor, Reinhold estimated the requirements listed in Table 2.2. Though fewer components would be needed for a smaller unit, the order of magnitude remains the same. These components must be built to precise specifications, in order to meet the high quality standards required in a nuclear plant. A government seeking to expand an established industrial base may hope that a nuclear project will stimulate demand and technological improvement. A government in a country with only rudimentary industrial infrastructure faces the more discouraging prospect of importing a larger share of the necessary components, draining foreign reserves and diverting scarce resources from the promotion of domestic industry.

When any nuclear program begins, extensive reliance upon foreign suppliers is unavoidable, even for the semi-industrialized countries, due to the extreme specialization of the requirements for nuclear stations. Large sums of capital must be raised for the purchase of foreign services and equipment. Since nuclear power stations cost more than $1 billion, officials in a country limited in foreign exchange reserves or heavily burdened by existing foreign debt – conditions widespread in the developing world – must think twice before increasing major financial commitments to foreign firms. In fact, the combination of tightened credit (raising dollar interest rates above the inflation rate) and the 100 percent increase in real oil prices following the Iranian revolution produced a major balance of payments crisis in the developing world, slowing down both borrowing and economic

Table 2.1 *Installed Electrical Capacity, 1978, Selected Countries (in MWe)*[a]

Africa		America		Far East	
Algeria	1,200	United States	602,008	Bangladesh	970
Libya	800	Argentina	10,275	Hong Kong	2,971
Morocco	980	Brazil	24,280	India	26,800
Egypt	3,944	Chile	2,925	Indonesia	1,660
Angola	523	Peru	2,580	South Korea	6,902
Zaire	1,217	Uruguay	850	Malaysia	1,575
Ghana	900	Colombia	3,775	Pakistan	2,236
Mozambique	1,397	Cuba	1,876	Philippines	3,760
Nigeria	960	Mexico	15,700	Thailand	2,820
Southern Rhodesia[b]	1,192	Venezuela	5,500	China	50,100
Tanzania	180	Total developing	43,010	Total developing	60,295
Zambia	1,710				
Total developing	19,121				

Middle East		Western Europe		Central Europe	
Israel	2,500	France	55,075	East Germany	17,958
Iran	5,300	Italy	23,200	Hungary	4,989
Iraq	860	West Germany	81,000	Poland	22,836
Kuwait	1,650	United Kingdom	72,000	USSR	247,687
Syria	940	Sweden	26,000		
Turkey	5,000	Greece	4,850	Oceania	
Total developing	12,783	Spain	28,350	Australia	22,823
		Yugoslavia	10,700	New Zealand	5,400
				Total developing	1,284

[a]Many figures are Statistical Office estimates.
[b]Now Zimbabwe.

Source: UN Department of International Economic and Social Affairs, Statistical Office, *World Energy Supplies 1973–1978* (New York: United Nations, 1979), pp. 246–60.

Table 2.2   *Mechanical and Electrical Components for Nuclear Power Plants with 1300 MWe PWR*

Component	Estimated quantity per unit
Heat exchanger	350
Tanks	200
Pumps and compressors	550
Valves	10,000
Cranes	25
Transformers	30
HV-motors	70
LV-motors	550
Special equipment	180

Source: H.-K. Reinhold, 'Transfer of technology', in *Problems Associated with the Export of Nuclear Power Plants* (Vienna: IAEA, 1978), p. 161.

growth. Table 2.3 illustrates the seriousness of the debt problem. For the countries listed in the table, the net oil imports and interest payments on debt rose from $2.1 billion in 1970 to $56.8 billion in 1980, or from 11.9 percent to 43.9 percent of the value of exports of goods and services.[5] Featured in the table are countries with both heavy

Table 2.3   *External Debt and Oil Imports, Selected Countries*

(1980 US$ billion)	Total external debt[a]	External debt owed to banks[b]	Gross interest payments Total	% Exports[c]	Net oil imports Amount	% Exports[c]
Brazil	61.2	46.3	7.0	32	10.0	45
Argentina	21.4	19.0	2.8	29	0.8	7
South Korea	27.0	15.7	2.5	11	5.6	25
Philippines	12.1	8.6	0.9	12	2.6	36
Chile	10.6	6.8	1.1	18	2.2	36
Thailand	7.0	5.1	0.6	6	2.7	27
Taiwan	7.5	5.6	0.8	4	4.5	20
Colombia	7.4	4.3	0.6	9	0.6	9
Turkey	19.9	3.7	1.1	21	2.9	55
Ivory Coast	4.1	3.0	0.4	12	0.3	9
Bolivia	2.6	1.5	0.2	20	0.0	0
India	18.5	1.0	0.4	4	6.3	58
Total	199.3	120.6	18.4	178	38.5	327

[a]All maturities estimated at year end.
[b]Bank for International Settlement data.
[c]Exports of goods and services.
Source: *Far Eastern Economic Review*, March 20, 1981, p. 47.

external debt and major nuclear power commitments: Brazil, Argentina, South Korea, the Philippines, Taiwan, and India. Although the size of these debts partly reflects nuclear investments already made, the continuing commitment to nuclear power in most of these countries shows that high foreign debt exposure does not always deter major nuclear power commitments.

COMPLEXITY

Nuclear power stations are among the most sophisticated devices devised by man. Advanced electronics guide their operation and gauge their performance. Rigorous specifications are enforced in component manufacture and installation. Enormous pressure vessels of steel several inches thick must be cast, and welding must be of high quality. The presence of radioactivity necessitates perfect seals and remote handling equipment. Automatically activated emergency core cooling systems and a host of other safety systems are mandatory. Many fuel cycle activities are also difficult and potentially dangerous: the handling of highly corrosive uranium hexafluoride gas, all of the uranium enrichment methods developed or under development, the manufacture of the zirconium alloy (zircalloy) tubes which hold the uranium fuel pellets, the encasement of highly radioactive wastes in an inert medium (vitrification or glassification), and the permanent disposal of the vitrified wastes in geological structures, where the possibility of leaching from the waste canisters will be minimized.

This complexity entails both costs and benefits. The most obvious cost is dependence on foreign suppliers. India is the only developing country which arguably has achieved nuclear self-sufficiency, though others are moving in that direction. Reliance upon foreign assistance in reactor construction or fuel supply increases a government's vulnerability to pressures from its industrial suppliers. Where the developing country begins to take over some of the difficult tasks for itself, the cost is translated from dependence to the diversion of funds and skilled manpower from other investments. A commercial nuclear power program may involve thousands of trained workers and technicians, in countries where skilled labor is scarce.

Complexity also increases risk, at two levels. At the level of the unit itself, the need for precise system interaction increases the number of things that can go wrong, while the reduced margin of error in a finely-tuned process increases the likelihood that they will. A box camera cannot perform as well as a Nikon, but it is less likely to break down and if it does it probably will be easier to fix. Nuclear reactors have often been unexpectedly shut down by system failures, so that many units have operated far below expected capacities. Complex systems may also increase the overall risk to society. Their full implications cannot be predicted, while the deleterious effects of entrenched technologies cannot easily be remedied. One analyst, David Collingridge, recalled how early concerns over the environmental impact of the automobile focused not upon engine emissions but rather upon the problem of dust, thrown up from untarred roads. He concluded that, 'By the time a technology is sufficiently well developed and diffused for its unwanted social consequences to become apparent, it is no longer easily controlled. Control may still be possible, but it has become very difficult, expensive, and slow.'[6] The measures a government is willing to take in response to these societal risks depends upon whether it views technology as a catalyst for progress or as a corrosive to traditional values and environmental quality. Many view serious concern with the social implications of technology as a luxury, affordable only by the already prosperous.

The potential benefits of the introduction of complex technologies, such as nuclear, center around developing country efforts to accelerate development and close the 'technological gap'. The introduction of the IAEA *Nuclear Power Planning Study for Indonesia* reflected a common if specious view in the developing countries toward the causes and significance of this gap:

> As is generally known, the gap between rich and poor countries originated in the nineteenth century when a number of nations missed the Industrial Revolution. This century has brought about innumerable scientific discoveries and revelations which have formed the basis for a rapidly advancing technology, causing initially a further widening in the gap, but in fact creating also the possibility for less developed countries to more or less catch [sic] up with the progress of the industrialized nations.

Planners hope that the technology learned in a nuclear power program could serve their countrymen well in other sectors requiring similar skills. For example, steel reinforcing concrete in reactor containment structures enables domestic industry to learn how to apply that technique to other construction ventures.

The benefit obtained depends on how well domestic industry can absorb the acquired technology and apply it to other sectors. Since most components unique to a nuclear station (such as the reactor core and instrumentation) are manufactured by foreign contractors, critics charge that as much or more technology can be transferred through nonnuclear electric projects. The turnkey contracts commonly used for nuclear stations reduce the technology transfer to mere operation and maintenance. Since many cooperation agreements include different technological packages, political authorities must decide which technologies most contribute to their development objectives.

Antagonists in the nuclear debate entertain entirely different notions toward the proper role of technology in development. Nuclear critics often believe that the technologies adopted in a developing country should *match* present capabilities. By this logic, since most developing economies have capital shortages and labor surpluses, investment in labor-intensive technologies represent the most efficient resource utilization, providing jobs and increasing labor productivity while conserving capital. A whole school of thought along these lines has developed, advocating the adoption of appropriate technologies, which are appropriate in that they (1) rely upon resources most available in a country, and (2) optimize societal goals, such as per capita income growth, full employment, or equitable distribution.[7] Underlying the appropriate technology approach is a desire to redress inequitable income distributions, to protect dominantly rural populations from exploitation at the hands of the small landed and industrial classes, and to avoid the unpleasant urban ghettos, pollution, and crass commercialism seen as inherent to Western style development.

Proponents of nuclear power, by contrast, often believe that the technologies selected should *surpass* present capabilities in order to stimulate growth. If only appropriate technologies are adopted, they argue, then development will proceed so slowly that the prosperity of the North will forever elude the South. To make the quantum leap required to 'catch up' in development, governments strive in advanced fields, which become enclaves of modernity from which (it is hoped) new skills will spill over into other sectors, promoting modernization throughout the economy. Ascending the 'learning curve' accelerates growth in productivity and income. Expanded gross national product improves conditions for all. (Appropriate technologists fear that wealth increases will largely be appropriated by the rich.) Short-term efficiency is sacrificed willingly to the goal of long-term prosperity. The sentiment resembles that expressed by President John Kennedy, who said that the United States chose to go to the moon not because it was easy but because it was hard, 'because that goal will serve to organize and measure the best of our energies and skills'.[8] Politically, this school prefers the maximum diffusion of high technology and the assimilation of the economic and social patterns characteristic of the industrialized world into the developing nations. The fecklessness of many nuclear policy debates is not surprising in light of the divergent social goals each side cherishes.

FUEL REQUIREMENTS

The most common nuclear fuel is uranium-235, obtained through enriching natural uranium. It is possible to split uranium-238 (which comprises over 99 percent of natural uranium) and thorium-232, but neither isotope is fissile; they cannot sustain chain reactions. Both, however, are known as fertile materials, because they are transformed to fissile isotopes when bombarded with neutrons: uranium-238 to plutonium-239, thorium-232 to uranium-233. Uranium-235, the only naturally occurring fissile isotope, occurs in such minute quantities that some day it may cease to be economically recoverable, at which time the fertile materials will have to be converted to fissile isotopes if nuclear power is to survive. All nuclear reactors not only split but also produce fissile atoms, through the conversion of fertile materials. The proportion of fuel converted to fuel consumed is described as the conversion ratio. Thermal reactors produce less fissile materials than they consume, so their conversion ratios are less than 1.0.

In one type of reactor, the fissile fuel (containing plutonium dioxide) is densely packed, surrounded by a blanket of fertile material, and bombarded by unmoderated, or 'fast', neutrons. More fissile material is produced than is consumed, for a conversion ratio over 1.0 (typically about 1.15). Such reactors are known as breeders, and liquid metal usually replaces water as the coolant. Experimental fast breeder reactors have been operated in France, the Soviet Union, Great Britain, and the United States. By multiplying by a factor of sixty the energy value of each kilogram of uranium, breeder reactors may extend the lifespan of nuclear power indefinitely.

Uranium attracts especially those countries which depend heavily upon imported oil, which have suffered inflation and reduced aggregate demand through the dramatic price increases of the 1970s. These governments wish to reduce oil imports, often through diversification of energy supplies. Where oil is used to generate electricity, cheaper methods are sought. Coal production in most places has deteriorated and will require years of investment to revive. Besides, coal is expensive to transport and its combustion pollutes the air. Hydropower resources are often dispersed and remote from consumption centers. Natural gas deposits may also be remote and difficult to transport. Geothermal, solar, tidal, and wind energy technologies are decades away from commercialization. These problems enhance the appeal of uranium. Many oil-exporting governments believe that they, too, should stop burning oil to generate electricity, considering the potential profits from petroleum exports. OPEC ministers expect their new wealth to generate higher rates of economic growth and therefore of electricity demand. Nuclear power appears a promising alternative;

some planners in these countries forecast that by the mid-1990s up to 40 or 50 percent of electricity will be nuclear in origin.

The price and availability of uranium and related fuel services influences both the economic and political acceptability of nuclear power. The price of uranium fell after 1959, due primarily to the saturation of military needs, and remained depressed through 1973. Over the next two years, uranium prices increased nearly fivefold in real terms, as a result of a complex combination of enrichment and reprocessing planning and contracting, rising oil prices, oligopolistic uranium producer behavior, and other factors.[9] (See Figure 2.1.) Since 1978, recession and reduced nuclear power demand have driven uranium prices back downward. Even large price fluctuations, however, only marginally alter the economics of nuclear power relative to its competitors. Nuclear power is so capital-intensive that a 100 percent increase in the price of uranium ore may add only 10 percent to the price of delivered electricity, whereas an equivalent increase in coal prices may add 35 percent to electricity prices for that fuel.[10]

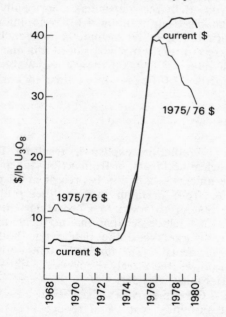

Figure 2.1   Uranium prices, spot deliveries 1968–80, as reported by Nuclear Exchange Corporation Quarterly Averages. 1975/76 $ are current dollars deflated by the general dollar GDP deflator for the entire OECD area. *Source*: Marian Radetzki, *Uranium; A Strategic Source of Energy* (London: Croom Helm, 1981), p. 14.

The politics of uranium are more pertinent. Around 63 percent of world uranium reserves reasonably assured at a price of up to $50/lb uranium oxide are found in Australia, Canada, the United States, and South Africa.[11] The first three of these have already restricted uranium exports on occasion in their efforts to prevent the spread of nuclear weapons. For obvious reasons, South Africa cannot necessarily be counted on to insulate uranium exports from political considerations. The historical monopoly enjoyed by the United States in providing uranium enrichment services helped persuade the Argentinians and Indians to build reactors using unenriched uranium. This monopoly has ended with the completion of uranium enrichment facilities in Europe and South Africa, and the entry of the Soviet Union to the international enrichment market. Uranium reserves in some developing countries, such as Niger and Gabon, may generate strong export earnings, but oil importers who lack uranium face the prospect of merely transferring their dependence on fuel imports from one group of countries to another. Technological dependence will probably remain more onerous than fuel dependence, especially as uranium becomes more available from uranium-rich developing nations. (See Table 2.4.) If thorium fuel cycles become commercially viable, the situation will be eased further, especially since India and Turkey have abundant thorium reserves: 649,000 tonnes reasonably assured, or approximately 60 percent of the presently estimated world total.[12]

SPECIAL RISKS

One special risk is radiation exposure, resulting from normal operation, from nuclear accidents, or from actions taken by criminals, terrorists, or governments. Radiation is present in all phases of the nuclear fuel cycle, from uranium mining, fuel production, and irradiation in the reactor to spent fuel processing and final waste disposal. The hazards 'include occupational accidents and radiation-induced disease in workers and the public due both to routine emissions and to accidents'.[13] The task here is not to evaluate the dangers of nuclear power, but merely to suggest the factors officials in developing countries must consider.

Risk is inevitable. Certain activities (like driving a car) increase risk while others (like fastening seat belts) reduce it. Risk cannot be eliminated from any activity. Consequently, assessments of the acceptability of risk cannot be made in isolation, on an absolute scale, but must be made in comparison to other activities. The risks from nuclear power can be evaluated in relation to the risks (1) from competitive power sources (coal pollutes the air, hydrodams break, gas

Table 2.4   *Uranium Resources by Continent[a] (in thousand tonnes)*

	Reasonably assured		Estimated additional	
	$80/kg U ($30/lb U_3O_8)	$130/kg U[b] ($50/lb U_3O_8)	$80/kg U ($30/lb U_3O_8)	$130/kg U[b] ($50/lb U_3O_8)
North America	**752**	**976**	**1,145**	**1,904**
USA	531	708	773	1,158
Canada	215	235	370	728
Mexico	6	6	2.4	2.4
Greenland	0	27	0	16
Africa	**609**	**776**	**139**	**263**
South Africa	247	391	54	139
Niger	160	160	53	53
Namibia	117	133	30	53
Algeria	28	28	0	5.5
Gabon	37	37	0	0
Central African Republic	18	18	0	0
Zaire	1.8	1.8	1.7	1.7
Somalia	0	6.6	0	3.4
Egypt	0	0	0	5
Madagascar	0	0	0	2
Botswana	0	0.4	0	0
Australia	**290**	**299**	**47**	**53**
Europe	**66**	**391**	**49**	**98**
France	39.6	55.3	26.2	46.2
Spain	9.8	9.8	8.5	8.5
Portugal	6.7	8.2	2.5	2.5
Yugoslavia	4.5	6.5	5	20.5
UK	0	0	0	7.4
Germany (F.R.)	4	4.5	7	7.5
Italy	0	1.2	0	2
Austria	1.8	1.8	0	0
Sweden	0	301	0	3
Finland	0	2.7	0	0.5
Asia	**40**	**46**	**1**	**24**
India	29.8	29.8	0.9	23.7
Japan	7.7	7.7	0	0
Turkey	2.4	3.9	0	0
Korea	0	4.4	0	0
Philippines	0.3	0.3	0	0
South America	**97**	**102**	**99**	**105**
Brazil	74.2	74.2	90.1	90.1
Argentina	23	28.1	3.8	9.1
Chile	0	0	5.1	5.1
Bolivia	0	0	0	0.5
Total (rounded)	1,850	2,590	1,480	2,450

[a]World outside communist areas.

[b]Includes resources at $80/kg U level.

Source: International Nuclear Fuel Cycle Evaluation, *Fuel and Heavy Water Availability: Report of Working Group 1* (Vienna: IAEA, 1980), p. 154.

tanks explode), (2) from unrelated activities (flying, smoking, swimming, lathe-operating), or (3) within the existing environment (cosmic radiation, cancer, living in New York City). At heart, risk acceptability is a nonscientific judgment, dependent on the values attached to various benefits (such as the mobility afforded by flight) and risks (such as the unpleasantness of dying tomorrow from a flood or in fifteen years from a radiation-induced cancer). Psychological effects can legitimately be included; the horror of an unlikely nuclear accident resulting in thousands of deaths may be weighed against the less dramatic certainty that each year a few people die from lung cancers induced by coal combustion.

In electricity generation, both health and environment are affected. Environmentally, coal is a serious offender. The carbon dioxide emitted in its combustion causes a 'greenhouse effect', a warming of the earth's climate with unknown but possibly grave long-term consequences. Coal mining takes over vast tracts of land, and water reacts with coal sulfur to form sulfuric acid which harms surrounding soil. Acid rain is another problem. Nuclear power entails mining risks similar to those for coal, but because far less uranium than coal is needed per megawatt of power, the effects are less severe. Natural gas pipeline leaks may ignite or suffocate surrounding flora and fauna. Hydro dams alter the whole surrounding ecosystem, as noted in the last chapter. Solar power may require that areas as large as those needed for coal mining, combustion, and ash disposal be covered by collecting plates.

Personal health effects also vary with energy source.[14] Occupationally, coal again appears to be the most dangerous, as illustrated in Table 2.5. Apart from accidents, coal miners are vulnerable to work-related respiratory ailments, or 'black lung'. Employees at uranium mines, mills, and reactors are exposed to more radiation than the public. With coal, public health is most affected by combustion-related air pollution, though studies conflict over the precise level of risk. With nuclear power, the most serious public radiation exposure arises from uranium mining and milling, and the reprocessing of spent reactor fuel. Exposure may cause cancers or genetic defects. Other major health risks stem from the possibility of major accidents – gas leaks or explosions, dam failures, nuclear plant leaks or core meltdowns. The environmental calculus is site-specific. In a small, crowded country like Taiwan, where air quality is already poor, nuclear power may appear much more ecologically benign than coal-fired plants.

Nuclear activities also present targets to terrorists or criminals. One approach is to steal weapons-grade uranium or plutonium and build a nuclear explosive. Another is to attack or occupy a nuclear facility and

Table 2.5 *Accidental Deaths, Injuries, and Workdays Lost During Routine Operations, by Energy Source (per gigawatt-plant-year)*

Energy Source	Accidental Deaths	Accidental Injuries	Workdays Lost[c]
Coal Mining[a]			
Mined underground	4.0	112	15,000
Surface mined	2.6	41	3,000
Oil	0.4	32	3,600
Gas	0.2	18	2,000
Nuclear[b]	0.2	15	1,500

[a]Synthetic liquid fuel from coal might be estimated to have a rate equal to that for coal plus an allowance for the conversion process.

[b]Table assumes once-through fuel cycle, without reprocessing. With reprocessing, the uranium oxide requirement could be reduced to 1.4 tons. Presumably, the mean extraction risk would be reduced proportionately, and the processing risk increased. The net result could be lower total risk.

[c]A permanently disabling accident was credited with 6000 workdays lost, and a temporary disability with 100 workdays lost. The figures are for 1977.

Source: Committee on Nuclear and Alternative Energy Systems (CONAES), *Energy in Transition 1985-2010* (San Francisco: W. H. Freeman, 1979), pp. 429-30.

threaten to destroy it or to trigger the release of large quantities of radiation. The difficulties associated with either approach are enormous. The radioactivity of the materials themselves protect against their misuse; the multiple safety systems of a reactor must be circumvented and physical security forces overcome. Still, an attack could succeed, if the attackers are well organized, skilled, and fearless. They could even obtain some advantage by merely *threatening* hostile actions, promising not to carry them out if certain demands were met. Some measures designed to prevent attacks, such as security screening of prospective employees and domestic surveillance on suspected terrorists, occasion concerns that civil liberties may suffer in the quest for nuclear safety, but such concerns have little currency among developing country governments.

Finally, nuclear programs may accelerate the spread of nuclear weapons to more governments. While it is true that there are easier, cheaper, more discreet paths to atomic bomb production than through civil research or power programs, no one can deny that such programs *may* be exploited in the manufacture of nuclear weapons. The technologies by which atoms are divided quickly to generate heat and very quickly to generate explosions cannot be neatly compartmentalized, isolating military from peaceful uses. India proved this in 1974, by using Canadian and American civil nuclear assistance to detonate a 'peaceful' nuclear explosive device. The possibilities of nuclear weapons proliferation, however, are unlikely to deter a government from beginning a nuclear power program, although they may discourage suppliers of nuclear technology from offering their wares to regimes with unstable or militant leaders.

Special risks demand special precautions. Regulatory bodies must rule that a project is safe before issuing sequential permits for site preparation, construction, fuel loading, and reactor operation. Regulatory personnel must be trained and guidelines promulgated. The burden can be eased by the adoption of existing nuclear regulations from the advanced nations. Inspections before and during operations are required for components, and radiation levels in the surrounding environment must be monitored. Regulation cannot be simplified; its usefulness inevitably owes much to the judgment of the regulators, who must make close calls on the acceptability of marginal variations from specification. Especially since nuclear power has become so politicized, physical protection for nuclear facilities must exceed that required for conventional stations.

HISTORICAL INTEREST IN ATOMIC ENERGY

The energy potential from fission was recognized from the outset of the atomic age. Indian nuclear efforts began in 1948, before those of many advanced nations, and Argentina entered the atomic energy field a year later. From 1945 to 1954, however, access to nuclear technology was constrained by the secrecy of the Americans, who refused to share their exclusive knowledge even with their wartime allies, let alone the newly emerging nations. This 'closed door' approach was abandoned in December 1953 by President Eisenhower in his Atoms for Peace proposals. By this time, the Soviet and French nuclear programs were already well underway, and it appeared impossible to arrest the spread of nuclear technology. The Atoms for Peace approach offered nuclear assistance to governments that promised to submit their nuclear programs to safeguards. In international conferences on the peaceful uses of atomic energy, held in Geneva in 1955 and 1958, large amounts of fission technology were declassified and disseminated to participants. The International Atomic Energy Agency, charged with assisting and regulating atomic programs, was created in 1957. During the 1950s and 1960s, the United States Government provided personnel training, research reactors, and fissile materials, often on a grant basis, to more than a score of developing countries.

By the early 1960s, nuclear power had entered into commercial use in North America and Europe, but remained out of reach for most developing countries, which found nuclear power neither economical compared to oil-fired stations, nor feasible for their small electrical grids. Because the supplies required for a nuclear program are often scarce in a developing country, electricity demand must be sufficient to merit large investments in their procurement. So long as the real price of oil was falling, in the 1950s and 1960s, demand for nuclear power could not induce a comprehensive effort. Indeed, by 1974 only three developing countries – India, Pakistan, and Argentina – had nuclear power plants in operation. In each of these countries strategic considerations decisively buttressed the economic arguments favoring nuclear development. Elsewhere, the investment was judged too burdensome.

Faced with enormous oil price increases in 1974, Third World leaders found the nuclear option to be more competitive, and plans to build reactors increased accordingly. According to the Atomic Industrial Forum, commercial generating units under construction, on order, or 'firmly planned' by 1976 had reached the levels of nine for Brazil, five for Egypt, five for India, three for Indonesia, five for Iran, ten for South Korea, nine for Mexico, three for Thailand, eight for Cuba, and two each for South Africa, Yugoslavia, Kuwait, and Libya.[15] Soon, however, this surge abated, as a consequence of several developments. Rising costs of nuclear power stations eroded their perceived economic advantage, while high inflation combined with recession sapped the ability of the oil-importing developing world to pay for reactors. Confidence in the reliability of some major nuclear suppliers waned, as countries such as the United States and Canada conditioned their nuclear exports upon increasingly stringent non-proliferation policy constraints. The 'demonstration effect' of the nuclear power programs in the advanced countries had always been important, but by the mid-1970s earlier demonstrations of the smooth introduction of reactors had been replaced by displays of effective opposition to nuclear power (especially in West Germany and the United States, two of the most aggressive nuclear exporters), and of the crippling effects of leaks and near accidents on nuclear programs (capped by the Three Mile Island incident). These deterrents to nuclear power programs emerged at the same time as development planners were increasingly urged not to emulate the environmental recklessness of the industrialized nations, and to consider carefully the possibility of alternative energy strategies, from the revival of coal production to the introduction of decentralized, renewable resource-based systems.

The Atoms for Peace approach was not universally admired even in the 1960s. Already some analysts feared that the diffusion of nuclear technology, safeguards notwithstanding, would facilitate the spread of nuclear weapons, and pointed to China as an example of a poor nation which nevertheless had joined the nuclear club. Increased concern over the further spread of nuclear weapons, following China's 1964 nuclear test, led to the successful conclusion in 1968 of the Non-Proliferation Treaty. Still the Atoms for Peace approach was enshrined in the Treaty, which offered 'the fullest possible exchange' of nuclear technology in exchange for the nonnuclear weapon states' pledge to forswear the nuclear option. Nevertheless, the 1974 Indian nuclear test fatally undermined the Atoms for Peace policy. The fear that India's test would be contagious provoked a drastic change in American nuclear policy. The new approach reached full expression under President Carter, who discouraged 'premature' entry into the breeder and plutonium economy and insisted that American atomic cooperation abroad be more strictly controlled, initiatives which incurred the rancor of America's allies as well as of the developing world generally.

The constraints on nuclear power development can only be overcome when both supply and demand are equal to the task. There must be sufficient supply of capital for construction, skilled manpower for operation and maintenance, and electrical grids for the transmission and distribution of the electricity to be generated. The supply of assistance – financial, technical, and industrial – from the advanced nations must be available as must political support ·for the nuclear export in question. There must be sufficient demand for base load electrical services, smokeless energy, and advanced technology. These requirements are rigorous, and all must be met simultaneously, a difficult feat in the erratic 1970s. OPEC stimulated but safety concerns depressed demand. Reduced domestic demand in the advanced nations increased the availability of supplies to developing countries, but increased nonproliferation concerns reduced supplier willingness to transfer certain nuclear technologies. The supplies within the developing countries could only be marshaled through concentrated effort, which was discouraged in the late 1970s by the reduced confidence that domestic demand and foreign supplies could sustain a major nuclear effort. By 1981, atomic energy advocates in developing countries hoped that relentlessly increasing oil prices, restored confidence in nuclear power, and the rekindling of American support for ambitious nuclear power targets would reestablish the confidence required to get their programs off the ground.

NOTES

1   I. Smart *et al., Report of the International Consultative Group on Nuclear Energy* (London: Royal Institute of International Affairs, 1980), p. 3.
2   Committee on Nuclear and Alternative Energy Systems (CONAES), *Energy in Transition 1985–2010* (San Francisco: W. H. Freeman, 1979), p. 276.

3 G. Woite, 'Can nuclear power be competitive in developing countries?', *Nuclear Engineering International* (July 1978), p. 46.
4 H.-K. Reinhold and H. Schweikert, 'Industrial infrastructure', in *Problems Associated with the Export of Nuclear Power Plants* (Vienna: IAEA, 1978), pp. 357–9.
5 *Far Eastern Economic Review* (March 20, 1981), pp. 46–7.
6 D. Collingridge, *The Social Control of Technology* (London: Frances Pinter, 1980), pp. 16–18.
7 A. Robinson (ed.), *Appropriate Technologies for Third World Development* (New York: St. Martin's, 1979), p. 140; and F. Stewart, *Technology and Underdevelopment* (London: Macmillan, 1977).
8 T. C. Sorensen, *Kennedy* (London: Hodder & Stoughton, 1965), p. 528.
9 M. Radetzki, *Uranium: A Strategic Source of Energy* (London: Croom Helm, 1981).
10 S. M. Keeny *et al.*, *Nuclear Power Issues and Choices* (Cambridge, Mass.: Ballinger, 1977), p. 126, illustrates the point for the American Midwest, but the general comparison applies elsewhere as well.
11 International Fuel Cycle Evaluation, *Report of Working Group I: Fuel and Heavy Water Availability* (Vienna: IAEA, 1980), p. 154.
12 Ibid., p. 198.
13 Keeny *et al.*, *Nuclear Power*, p. 185.
14 This section draws heavily from the CONAES study, *Energy in Transition*, op. cit., pp. 48–61.
15 Atomic Industrial Forum News Release, June 2, 1976, in Congressional Research Service, Library of Congress, *Nuclear Proliferation Factbook* (Washington, D.C.: US Government Printing Office, 1977), pp. 239–48.

Part Two: Policies

3 Three Approaches

Their strategic importance guarantees that nuclear policies will be directed from the highest government levels. These decisions are not made in a vacuum, wherein each advantage and disadvantage is assigned a value, and the summation of pluses and minuses determines the outcome. This is not only because it is impossible to assign specific values (or even general values on a single scale) to the incomparable, often psychological, effects of a nuclear program, but also because the 'vacuum' prerequisite to performing such a calculation does not exist and cannot easily be simulated. Domestic political infighting and foreign pressures could undermine the evenhandedness of any weighing of pros and cons. In order to assess how, in fact, nuclear policies are determined, it is necessary to analyze the different approaches to nuclear power adopted in developing nations.

Developing country governments may be grouped into three categories:

(1) those that have not introduced nuclear power to their countries,
(2) those that pursue nuclear power programs with high priority on developing self-sufficiency and reducing dependence on the advanced nations, and
(3) those that pursue nuclear power programs with less concern for independence and more for bringing large quantities of nuclear power on line quickly.

The first category includes most of the developing world. The distinction between the second and third is sometimes blurred, as when a government orders several reactors *and* the ancillary fuel facilities which promote self-sufficiency (Brazil), or orders reactors from abroad while independently carrying out its own nuclear fuel production efforts (Pakistan). Nevertheless, India and Argentina may confidently be placed in the class of those most concerned with self-sufficiency. Israel, Pakistan, and South Africa round out this group. Governments eager to exploit fission for electricity generation but content to rely on foreign support in this effort include Egypt, Mexico, the Philippines, South Korea, Taiwan, Yugoslavia, and, formerly, Iran.

Economic circumstances alone do not dictate nuclear policies. A relatively wealthy developing country, like Venezuela, may contemplate nuclear power far less seriously than those more deprived, like Bangladesh or Pakistan. Oil situations give no sure indicator. Some oil-rich countries (Iran, Mexico, Argentina) have begun nuclear power plant construction; others (Venezuela, Nigeria, Kuwait) have not. Although oil-poor nations display a natural interest in nuclear power, only a few have taken up the option. One could run down the list of indicators – uranium resources, technological capability, capital availability – without finding any that cleave the nuclear from the nonnuclear nations. Consequently, in order to make some sense out of the possible motives which induce some governments to accept and others to reject nuclear power, it is first useful to distinguish between the broad approaches they take.

NONNUCLEAR COUNTRIES

Too little attention has been devoted to nuclear policy-making in governments which have either rejected, postponed, or simply not yet adopted the nuclear option. Though it is natural to focus nuclear policy studies on countries with nuclear plans, it is also misleading; the newsworthiness of these countries' programs sometimes conveys an exaggerated impression that there is a general clamor for nuclear power in the developing world, stemmed only by the advanced nations' reluctance to transfer technology. In fact, the difficulties of nuclear power development have greatly influenced most developing countries. Admittedly, the impoverished nations of Asia and black Africa hardly have a nuclear option to reject (though a power reactor has been under consideration in Bangladesh since at least 1963).[1] Others, though, are in many ways comparable to the nuclear developing countries. In Latin America, the country with the highest *per capita* income, Venezuela, has been joined by Chile, Colombia, and others in abstaining from nuclear power. The most populous African nation, Nigeria, has followed suit, while outside Egypt nuclear efforts in the Magreb have never exceeded the desultory. This is also true in four of the five Asean nations (Indonesia, Malaysia, Singapore, and Thailand) and in the Middle East, where Saudi and Kuwaiti proposals for large nuclear complexes to drive desalination plants and other industries have borne no fruit. The Turkish pattern is typical: occasional eruptions of interest in buying a reactor, followed by snail-paced negotiations, perhaps a signature on a letter of intent, further haggling over prices and credit arrangements, and so on. Often by this time the initial impetus behind the project has dissipated, a victim of the perennial financial crises these countries endure. In 1955, Turkey began a nuclear collaboration with the United States and six years later inaugurated its first research reactor. By 1976, Swiss architect-engineers were on hand, and the Turkish Electrical Authority had received a license for a 600 MWe reactor. Feasibility studies followed, along with negotiations with Italian, Swedish, British, American, French, and German firms. The preferred offer varied with the political sympathies of the government in power. In 1979, the Soviet Union reportedly agreed to build two 440 MWe reactors on the Black Sea coast for around $880 million. The replacement of Bulent Ecevit by a more conservative prime minister, Suleyman Demirel, scotched that proposal. Then, talks with Asea-Atom reportedly stalled due to Turkish insistence that Sweden provide a 100 percent guaranteed credit for the $1 billion plus project, instead of the 85 percent offered. In 1981, another one-year delay was announced. As of this writing, no contract had been awarded.

The pure case against nuclear power in these countries can be adduced simply by condemning its characteristics; it is too capital intensive, too complex, too dangerous, and too uranium-dependent. That this argument is persuasive in most developing countries is not surprising. Many already face serious foreign debts and foreign exchange shortages, and look askance at any project that will require massive payments to foreign contractors. The large scale which underpins the competitiveness of fission-generated electricity also deters the developing countries, which lack the grid capacity to support large units. The rule of thumb that one unit should not contribute more than 10 percent of an electrical grid's capacity implies that a 600 MWe reactor, the smallest size commonly offered, must be placed in a system of at least 6,000 MWe capacity, which excludes the vast majority of developing countries. The complexity of nuclear technology places a premium on one of the scarcest Third World resources – skilled manpower – while threatening to create enclaves so highly developed that few if any other sectors will be advanced enough to benefit from the technology acquired. In countries lacking adequate uranium reserves, the volatility of the nuclear fuel market can deter investment.

Environmental considerations exist, but are muted. Traditionally, developing countries have tended to view environmental protection as a luxury, affordable only by nations wealthy enough to divert resources from goods production to conserving clean air and water. Recently this view has been revised, for two reasons. First, in an age of increasing energy prices, economic and environmental interests have become more coincident. Protecting forests also preserves valuable firewood supplies and prevents declining crop yields caused by soil erosion. Using agricultural and animal wastes for fuel may save money which otherwise would be spent on oil purchases or coal production. The environmental lessons of the Three Mile Island (TMI) accident might be debatable, but it undoubtedly cost General Public Utility hundreds of millions of dollars in substitute electricity purchases from neighboring utilities, while leaving a $700 million reactor standing idle as a wasting asset for at least three years.

Second, environmental concerns increasingly merge with political concerns. Where environmental issues attract widespread popular support, a government can ignore them only at its peril. Fragile governments may lose domestic support to critics articulate enough to exploit public fears. The Marcos government provides a good example. The Kalingas tribe displaced by the Chico river basin hydro development project sparked a major controversy by murdering government officials. President Marcos, sensitized to such concerns, suspended construction of a 600 MWe reactor following the TMI accident, pending the results of an inquiry into the ability of the Philippine government to cope with a similar incident. (Work there later resumed). The magnitude of property damage and personal injury possible in a nuclear accident are too great to be ignored, by any government. Even if a buyer is incautious, the nuclear suppliers cannot be, due to concerns for safety, commercial interests in avoiding business-damaging accidents, and government regulations. The tangible results of environmental concerns are the limitation of reactor sites to what are judged to be geologically stable areas, the imposition of stringent Western safety specifications to developing country reactor projects, and a consequent increase in the price of nuclear stations. Though countries can sometimes avoid subjection to additional safety regulations from the exporting governments, after a reactor project is already under way, many safety features are non-discretionary, regardless of the importing government's ecological sensitivities.

The burden of uranium dependence cannot be viewed in isolation. The burden of *any* dependence cannot easily be quantified, since it hinges upon calculation of amorphous factors: the likelihood that any supplier will suddenly impose political conditions on or terminate supplies, or raise prices to unpredictable levels; the possibility that the offending supplier will be joined by others in its caprice; the probable evolution of availability, prices, and lead times for various substitutes. Theoretically, approximations for all these variables can be incorporated into economic choices among alternatives. In reality, the variables are so many and their approximations so encrusted by uncertainty, that political judgment becomes the only plausible recourse.

The uranium (plus technological) dependence implied by nuclear programs is weighed against oil dependence or, more importantly for the long run, against the dependence entailed in alternative energy strategies. In this comparison, domestic fuel resources play a special role. Those sitting atop plentiful oil reserves may display either little tolerance for or little concern over uranium dependence. Nigeria and Malaysia fit the former category; Mexico, the latter. Taiwan, where dependence on oil imports will increase from 80 to 89 percent of total energy supplies in the coming decade even *with* an ambitious nuclear program, finds uranium dependence a desirable partial substitute for oil dependence. Thus, natural energy resource endowments powerfully affect nuclear policy decisions, but in unpredictable directions. In poorly endowed countries, which cannot avoid energy dependence, uranium may seem to present no more onerous vulnerability than oil. Another effect of mineral poverty, however, may be to deprive a government of the wealth needed to initiate a nuclear program. Better endowed governments have less need to subject themselves to the risk of uranium dependence, but also less need to fear such dependence, and greater financial ability to embark upon the nuclear path.

Since resource endowments can cut both ways, they do not identify those governments most likely to regard the nuclear option skeptically as well as do capital shortages, small grids, and skilled labor shortages. Apart from these obvious features, the nonnuclear countries often possess one other trait: unglamorous ambitions. The term 'glamorous' is used here neutrally, connoting neither approval nor disapproval. The important distinction highlights those governments willing to devote large sums to assimilating high technology as a measure of status. Nonnuclear governments on the whole are less inclined to spend heavily on showy projects, sometimes fearing that they might appear profligate and lose public support, sometimes simply more concerned with the basic tasks of supplying their citizens with food, water, education, health care, irrigation, and roads. Their ambitions to become major regional or eventual world powers are either weak or underfinanced, and their notion of the importance of 'status' and what activities do or do not contribute to it vary widely.

Lack of nuclear power construction does not imply a lack of interest in nuclear technology. Many developing country governments have supported nuclear research, many possess atomic energy commissions which eagerly promote nuclear power, and many, at times, have expressed 'firm plans' to order commercial power reactors. Often, atomic energy commissions have lobbied vigorously for the adoption of a nuclear power program, but have not prevailed over the bureaucratic competition. This has been true in Indonesia, Turkey, and Thailand. To be sure, bureaucratic defeats can be reversed. If rising energy costs for alternatives inflict increasing economic losses, and the crisis of confidence in nuclear power passes, reactor plans in developing countries may revive. Due to its long lead time, however, even if desires for nuclear power surged immediately, it could not make an important contribution to electricity supply until the 1990s. The governments of the South can defer the nuclear option without penalty, for the nuclear industries of the North will be happy to export their wares as long as they are able. By contrast, when the US Congress terminated the supersonic transport project, there was no other government or entrepreneur in the wings to come to the rescue. The decision could only be reversed at the enormous cost of reassembling the dispersed staff and other elements of a crumbled effort. Developing country governments for the foreseeable future will have the luxury to vacillate at little cost, and proceed with a nuclear power scheme only when they view the conditions as ideal. In the meantime, they may continue the relatively inexpensive experimental and training programs in order to ease the job of introducing nuclear power, should they decide to do so.

In fact, many developing countries did initiate nuclear research efforts in the 1950s and 1960s, mostly through American assistance. The US Atomic Energy Commission and Agency for International Development provided grants, personnel training, and supplies of sensitive materials to over a score of developing countries, including those listed in Table 3.1. By 1975, research reactors could be found throughout Latin America (Argentina, Brazil, Chile, Colombia, Mexico, Uruguay, Venezuela), the Middle East (Egypt, Iraq, Iran, Israel, Turkey), and Asia (India, Indonesia, South Korea, Philippines, Thailand). (See Table 3.2.) Atomic research programs can be cheap – a simple, zero-power reactor may cost a few hundred thousand dollars – and it made sense to governments in these countries to undertake such modest investments in a technology that could bring great economic benefit in the future.

One other, more sinister, motive is attributed to developing countries' research efforts: the desire to be able to build nuclear weapons. Small research reactors are more dangerous in this regard than are large power reactors; first, because they can be built without foreign assistance and therefore without safeguards, and second, because they often use highly enriched uranium (90 percent uranium-235), which can be used for nuclear explosives. Power reactors now available use uranium enriched to only 3 or 4 percent uranium-235. Plutonium, another weapons-usable fissile substance, can also be obtained from the reprocessing of research reactor spent fuel. Plutonium reprocessing can be mastered more easily than uranium enrichment, so countries unable to import highly enriched uranium need not despair of manufacturing nuclear explosives. In fact, an unsafeguarded Indian facility reprocessed spent fuel, irradiated in a Canadian-American-supplied research reactor, CIRUS, to provide the plutonium used in the May 1974 test at Pokharan. Unsafeguarded, small-scale reprocessing laboratories were built and dismantled in Argentina and Taiwan, while Israel is also assumed to have reprocessing capability. Concerns that 'peaceful' nuclear research may be misused may persist even when a country joins the Non-Proliferation Treaty and accepts international safeguards. Iraq is the prime example. The French agreed to supply Iraq with a 70 MW research reactor, Osirak, which uses 90 percent enriched uranium fuel. After the core of the Osirak was destroyed by a mysterious explosion while it sat awaiting shipment in the southern French port of La Seyne-sur-Mer, the French took the opportunity to urge acceptance of a substitute design requiring lower enriched uranium. Baghdad demurred, insisting that Paris honor its commitment to supply the weapons-usable uranium originally offered. The French agreed, noting that Iraq was an NPT-party and insisting with IAEA support that the material was adequately safeguarded.

The Iraqis also proceeded in 1977 to buy four laboratories for some $500,000 from Italy, including a 'hot-cell' laboratory, which enables technicians to handle irradiated fuel for plutonium reprocessing. The 1979 agreement for cooperation with Brazil sparked surmise that the Iraqis sought indirect access to the enrichment and reprocessing technologies promised by West Germany to Brazil. Finally, reports alleged that the Iraqis had quietly acquired from Brazil a stockpile of natural uranium, which converts to fissile plutonium when irradiated. Weighed against the lack of an active atomic power program, this keen interest in suspicious activities was so premature that it justifiably provoked grave concern worldwide. Acting on that concern, on June 7, 1981 Israeli jets bombed the Iraqi research reactor complex, leaving one French technician dead and the nearly completed complex in irreparable wreckage.

THE NUCLEAR COUNTRIES

The standard case for nuclear power in developing countries proceeds from the premise that since energy demand grows in step with national income, energy supply must increase rapidly to sustain rapid income growth. This theory also holds that integration of the national market and replacement of barter by a monied economy are essential to

Table 3.1 Atoms-for-Peace Program Grants for Research Reactors, Fiscal Years 1956–62

Country	Power	Manufacturer	Estimated project cost ($m.)	Amount ($)	Fiscal year awarded
Argentina	5 MW	Argentine National Atomic Energy Commission	–	350,000	1962
Austria	5 MW	American Machine & Foundry	4.0	350,000	1958
Belgium	25 MW	Centre d'Etudes de l'Energie Nucléaire	10.01	350,000	1958
Brazil	5 MW	Babcock and Wilcox	1.3	350,000	1956
China (Taiwan)	1 MW	International General Electric	1.0	350,000	1958
Colombia	10 kW	American Machine & Foundry	–	350,000	1962
Denmark	5 MW	Foster-Wheeler	1.4	350,000	1956
Greece	1 MW	American Machine & Foundry	1.3	350,000	1958
Indonesia	100 kW	General Atomic	0.8	350,000	1961
Iran	5 MW	American Machine & Foundry	4.8	350,000	1962
Israel	1 MW	American Machine & Foundry	1.4	350,000	1958
Italy	5 MW	American Car & Foundry	3.6	350,000	1958
Japan	10 MW	American Machine & Foundry	1.5	350,000	1957
Korea, South	100 kW	General Atomic	1.1	350,000	1959
Netherlands	20 kW	American Car & Foundry	3.9	350,000	1956
Norway	10 kW	Norstom	0.8	350,000	1958
Pakistan	5 MW	American Machine & Foundry	3.5	350,000	1960
Portugal	1 MW	American Machine & Foundry	1.0	350,000	1957
Spain	3 MW	International General Electric	1.0	350,000	1956

Sweden	30 MW	American Car & Foundry	4.3	350,000	1958
Thailand	1 MW	Curtiss-Wright	0.82	350,000	1959
Turkey	1 MW	American Machine & Foundry	2.88	350,000	1960
Venezuela	3 MW	International General Electric	5.0	350,000	1957
Vietnam	100 kW	General Atomic	0.75	350,000	1959
West Germany	1 MW	American Machine & Foundry	3.1	350,000	1958
Yugoslavia	100 kW	General Atomic	—	200,000	1961
Total				$8,950,000	

Source: US Comptroller General, 'US financial assistance in the development of foreign nuclear energy programs', Report ID-75-63, May 28, 1975, p. 60.

Table 3.2   *Research Reactors and Critical Assemblies under IAEA Safeguards or Containing Safeguarded Material, December 31, 1979*[a]

Country	Abbreviated name	Location	Type	Capacity MW(th)
Argentina	RA-1	Constituyentes	Tank	0.07
	RA-2	Constituyentes	MTR	0.00
	RA-3	Ezeiza	MTR	5.00
	RA-4	Rosario	SUR-100	0.00
Brazil	IEAR-1	São Paulo	MTR	5.00
	UMG	Belo Horizonte	Triga I	0.10
	RIEN-1	Rio de Janeiro	Argonaut	0.01
Colombia	IAN-R1	Bogota	MTR	0.02
Chile	La Reina	Santiago	Herald	5.00
	Lo Aguirre	Santiago	MTR	10.00
Indonesia	PRAB	Bandung	Triga II	1.00
	Gama	Jogjakarta	Triga II	0.25
Iran	TSPRR	Tehran	Pool	5.00
Iraq	IRT-2000	Baghdad Tuwaitha	Pool	2.00
Israel	IRR-1	Soreq	Pool	5.00
South Korea	KRR-TRIGA II	Seoul	Triga II	0.10
	KRR-TRIGA III	Seoul	Triga III	2.00
Mexico	Centro Nuclear de Mexico	Ocoyoacac	Triga III	1.00
	Training reactors facility	Mexico City	SUR 100	0.00
Pakistan	PARR	Rawalpindi	Pool	5.00
Peru	RP-O	Lima	Tank	0.00
Philippines	PRR-1	Diliman, Quezon City	Pool	1.00
South Africa	SAFARI-1	Pelindaba	Tank	20.00
Thailand	TRR-1	Bangkok	Pool	2.00
Turkey	TR-1	Istanbul	Pool	1.00
	TR-2	Istanbul	Triga II	0.25
Uruguay	RU-1	Montevideo	Lockheed	0.10
Venezuela	RVI	Altos de Pipe	Pool	3.00
Yugoslavia	Triga II	Ljubljana	Triga II	0.25
	Boris Kidric R.	Vinča	Tank	6.50
	RB	Vinča	Critical assembly	0.00
Zaire	Triga-Zaire	Kinshasa	Triga II	1.00

[a]For unsafeguarded facilities, see Table 4.1 of IAEA, *The Annual Report for 1979* (Vienna: IAEA, 1980), pp. 47–50.

development. Accordingly, planners seek to commercialize the energy sector, most visibly through rural electrification schemes. Important advantages achieved in the transition from traditional to more sophisticated energy use patterns include relief of peasant abuses of over-harvested forests, reduced fuel waste by the introduction of more efficient stoves, and better use of biomass either by burning or fermentation into carbon fuels.

Electricity plays a special role, symbolically and economically. Widespread electrification symbolizes technical progress and national integration. Electricity transports heat in a highly concentrated and flexible form. It can be used in tasks ranging from the lowest to highest grades, from space-heating to motor drives to arc welding. Not surprisingly, to deliver energy in such a versatile form is costly. Three watts of heat are required to generate one watt of electricity. This one-to-three ratio means that electricity has a thermal efficiency of roughly 33 percent. A further 10 percent of the electricity generated may be lost in its transmission through cables and wires.

Electricity may be generated in several ways, through wood or fossil fuel combustion, nuclear, geothermal, or hydro stations, photovoltaic cells, windmills, or other unconventional technologies. Proponents of fission power naturally find the alternatives sorely lacking. Oil is expensive. Coal is dirty and difficult to mine and transport. Hydro resources are often remote from consumption centers and need even more capital investment for exploitation than do nuclear. Collecting and transporting natural gas is difficult and expensive, and many wish to conserve it for petrochemical uses. Geothermal resources may also be remote and insufficient to anticipated demand, given present technologies. Wood or charcoal could be used, but at the moment it is more important to restrain pressure upon their supplies. Biomass and solar technologies are so far from commercialization that they are still known as 'exotic' technologies.

The paramount advantage of nuclear power is its cost. The cost differential cannot be quantified generically, because it varies by country, according to land, labor, and capital availability for various technologies. In general, where its low lifetime fuel costs outweigh the large initial outlays, nuclear power becomes economical. These factors will be more fully considered in Chapter 8. The pure case for nuclear also argues that it is a well-established technology, after over two decades of commercial use, and that its large scale will accelerate poorer countries' efforts to improve their position relative to the North. The environmental and health risks of nuclear power are judged acceptable, especially when viewed in light of the severe air pollution which already defiles crowded urban areas in developing countries. Where possible, fission can take advantage of domestic uranium and eventually thorium deposits. When the cost differential compared to alternatives is marginal, an important concomitant advantage for fission is that it builds up a country's technological base. Also, if the cost differential marginally favors nuclear power or even decisively favors alternatives, a popular strategy is to forecast energy demand increases so vast that *all* feasible alternatives (including nuclear) must be deployed, if growth is not to suffer.

This pure case applies generally to countries already embarked on the nuclear path. Although every case is unique, enough similarities exist to draw one major distinction, between the 'independent' and 'dependent' approaches toward nuclear power development. These adjectives are only relative. No developing country has ever been able to carry out a truly independent nuclear program. Only the governments of the United States, Canada, Great Britain, France, and the Soviet Union can claim that achievement. Of these, only the United States, Canada, and the Soviet Union are self-sufficient; the others depend on uranium imports. Also, France abandoned its own gas–graphite reactor in 1969 in favor of the Westinghouse pressurized water reactor design, and Great Britain may yet follow suit, having encountered serious problems in seeking to develop a successor to its first-generation gas–graphite reactor, the Magnox. Most European nations and Japan have proceeded as far, or nearly so, in nuclear capabilities. Only after this large group do we reach even the most advanced developing countries. Nuclear plants operable, under construction, or on order in the developing world are shown in Table 3.3.

THE INDEPENDENTS

Then why use the term 'independent' at all? It *does* usefully distinguish between significantly different approaches to nuclear power development. A government may view atomic energy as a potentially important contributor to economic development, but one whose development should be guided by the same criteria as any other activity, to be introduced only when its costs and net benefit compare favorably to competitive alternative activities. In this view, noneconomic factors, such as prestige or political leverage, are not decisive. Alternatively, atomic energy may be viewed symbolically, not merely as an engine of growth, but also as a litmus of independence, sophistication, or status, a technology whose mastery divides strong from weak, rich from poor, exploiters from exploited. Governments sympathetic to this perspective sometimes subordinate economic evaluation criteria to political goals. This politicized approach is relatively 'independent', in that governments which adopt it proclaim the importance of their

Table 3.3 Nuclear Power Plants Operable, under Construction, or on Order (30 MWe and Over) as of December 31, 1981, Selected Countries

	Net MWe	Type	Reactor supplier	Generator supplier	Architect engineer	Constructor	Construction stage (%)	Commercial operation Orig. schedule	Actual or expected
Argentina									
Comisión Nacional de Energía Atomica[a]									
Atucha 1 (Lima, Buenos Aires)[a]	335	PHWR	Siemens	KWU	Siemens	Siemens/Imp.	100	6/72	6/74
Atucha 2 (Lima, Buenos Aires)	692	PHWR	KWU	KWU	CNEA/KWU	CNEA/KWU	3		indef.
Embalse (Embalse, Rio Tercero)	600	PHWR	AECL	Italimpianti	AECL/Italimpianti	Italimpianti/AECL	85	12/79	early 83
Brazil									
Furnas									
Angra 1 (Itaorna)	626	PWR	W	W	G&H/PE	W	100	3/77	7/82
Angra 2 (Itaorna)	1245	PWR	KWU	KWU	Nuclen	KWU	10	12/82	6/87
Angra 3 (Itaorna)	1245	PWR	KWU	KWU	Nuclen	KWU	1	6/84	12/88
Egypt									
Egyptian Electricity Authority									
Sidi-Krier-1 (Sidi Krier)	622	PWR	W	W	Gilbert	Jones	0	/83	/88
India									
Atomic Energy Commission, Department of Atomic Energy									
Tarapur 1 (Bombay)[a]	200	BWR	GE	GE	Bechtel	Bechtel	100	2/69	10/69
Tarapur 2 (Bombay)[a]	200	BWR	GE	GE	Bechtel	Bechtel	100	2/69	10/69
RAPP 1 (Kota, Rajasthan)[a]	202	PHWR	CGE	EEC	AECL/MECO	HCC	100	12/69	12/73
RAPP 2 (Kota, Rajasthan)[a]	202	PHWR	L&T	EEC	AECL/MECO	HCC	100	12/73	4/81
MAPP 1 (Kalpakkam, Tamil Nadu)[a]	220	PHWR	L&T	BHE	DAE	ECC	99.2	6/76	/82

Table 3.3 Nuclear Power Plants Operable, under Construction, or on Order (30 MWe and Over) as of December 31, 1981 (Contd.)

	MWe	Type					%		
MAPP 2 (Kalpakkam, Tamil Nadu)	220	PHWR	BHE	L&T	DAE	ECC	81	6/77	/84
NAPP 1 (Narora, Uttar Pradesh)	220	PHWR	BHE	WIL	DAE	HCC	57	3/81	/86
NAPP 2 (Narora, Uttar Pradesh)	220	PHWR	BHE	R&C	DAE	HCC	18	3/82	/87
New Project 1 (site undecided)	220	PHWR							/90
New Project 2 (site undecided)	220	PHWR							/91
Iraq									
Iraq 1	900	PWR					0		
Korea									
Korea Electric Co.									
Korea Nuclear 1 (Ko-Ri, near Pusan City)[a]	564	PWR	GEC	W	Gilbert	W	100	12/75	4/78
Korea Nuclear 2 (Ko-Ri, near Pusan City)	605	PWR	GEC	W	Gilbert	W	81.1	11/79	2/83
Korea Nuclear 5 (Ko-Ri, near Pusan City)	900	PWR	GEC	W	Bechtel	Utility	69.7	9/84	9/84
Korea Nuclear 6 (Ko-Ri, near Pusan City)	900	PWR	GEC	W	Bechtel	Utility	54.5	9/85	9/85
Korea Nuclear 7 (Young Kwang-Kun)	950	PWR	W	W	Bechtel	Utility	16.8	3/86	3/86
Korea Nuclear 8 (Young Kwang-Kun)	950	PWR	W	W	Bechtel	Utility	14.8	3/87	3/87
Korea Nuclear 9 (UI Jin-Kun, near Yeongju City)	950	PWR	undecided	Fra	Fra	Utility	5.9	12/87	3/88
Korea Nuclear 10 (UI Jin-Kun, near Yeongju City)	950	PWR	undecided	Fra	Fra	Utility	5.9	12/88	3/89
Korea Nuclear 3 (Wolsung-Kun)	629	PHWR	Parsons	AECL	AECL/Canatom/AC	AECL	93	1/82	12/82
Libya									
Libya 1	300	PWR	AEE	AEE			0	indef.	
Mexico									
Comision Federal de Electricidad (CFE)									
Laguna Verde 1 (Laguna Verde, Veracruz)	654	BWR	Mitsubishi	GE	Ebasco	CFE/Ebasco	67	6/77	1/84
Laguna Verde 2 (Laguna Verde, Veracruz)	654	BWR	Mitsubishi	GE	Ebasco	CFE/Ebasco	35	6/78	1/85

Pakistan

Pakistan Atomic Energy Commission

Station	Capacity	Type					%	Start	Op.
Kanupp (near Karachi)[a]	125	PHWR	CGE	Hitachi	CGE	CGE	100	6/71	12/72

Philippines

Philippine National Power Corp.

Station	Capacity	Type					%	Start	Op.
PNPP 1 (Morong, Bataan Luzon)	620	PWR	W	W	B&R	W	23	12/82	/85

South Africa

Electricity Supply Commission (ESCOM)

Station	Capacity	Type					%	Start	Op.
Koeberg 1 (Koeberg)	922	PWR	Fra	Alsthom		SB	70	12/82	1/83
Koeberg 2 (Koeberg)	922	PWR	Fra	Alsthom		SB	50	12/83	1/84

Taiwan

Taiwan Power Co.

Station	Capacity	Type					%	Start	Op.
Chin-shan 1 (Shihmin Hsiang)[a]	604	BWR	GE	W	Ebasco	TPC	100	12/75	12/78
Chin-shan 2 (Shihmin Hsiang)[a]	604	BWR	GE	W	Ebasco	TPC	100	12/76	9/79
Kuosheng 1 (Kuosheng)[a]	951	BWR	GE	W	Bechtel	TPC	100	10/78	12/81
Kuosheng 2 (Kuosheng)	951	BWR	GE	W	Bechtel	TPC	96	10/79	10/82
Maanshan 1 (Maanshan)	907	PWR	W	GE	Bechtel	TPC	46	4/81	4/84
Maanshan 2 (Maanshan)	907	PWR	W	GE	Bechtel	TPC	25	4/82	4/85

Turkey

Station	Capacity	Type					%	Start	Op.
Akkuyu (Akkuyu)	440	PWR	AEE				0		/87

Yugoslavia

Savske Electrane (Slovenia) and Elektroprivreda (Croatia)

Station	Capacity	Type					%	Start	Op.
Krsko (Krsko)[a]	615	PWR	W	W	Gilbert	W/local	100	12/78	12/81

[a]Units in commercial operation.

Abbreviations Used in Table

AC	Acres Canatom (Canada)	G&H	Gibbs & Hill Inc. (USA)
AEE	Atomenergoexport (USSR) (formerly TPE Technoprom-export)	Gilbert	Gilbert Associates Inc. (USA)
		HCC	Hindustan Construction Co. (India)
Alsthom	Sté. Generale de Constructions Electriques et Mechaniques (France)	Imp	Impresif
		Jones	J. A. Jones Construction Co. (USA)
Bech	Bechtel Corporation (USA)	L&T	Larsen & Toubro (India)
BHE	Bharat Heavy Electrical (India)	MECO	Montreal Engineering Co. (Canada)
BWR	Boiling water reactor	PE	Promon Engenharia SA (Brazil)
CFE	Cie d'Enterprises CFE SA (Belgium)	PHWR	Pressurized heavy water moderated and cooled reactor
CGE	Canadian General Electric	R&C	Richardson & Cruddas (I) (India)
Ebasco	Ebasco Services Inc. (USA)	SB	Spie Batignolles SA (France)
EEC	English Electric Co. Ltd (Canada)	TPC	Taiwan Power Company
Fra	Framatome Societé Franco-Américaine de Constructions Atomiques SA (France)	W	Westinghouse Electric Corporation (USA)
		WIL	Walchandnagar Industries Ltd (India)
GE	General Electric Co. (USA)		

Source: Nuclear News (February 1982), pp. 83-102.

independence and the contribution nuclear power makes to it. The independent approach entails some expense, both in opportunity cost for resources diverted from more productive but less glamorous uses, and in the inefficiency unavoidable when a complex technology is introduced prematurely. Proponents of the independent approach reply that short run opportunity cost and inefficiency are outweighed in the long run by the 'leapfrogging' nuclear power permits to higher levels of productivity and development, but their argument must remain speculative until time casts its verdict.

Defining the term 'independent' does not solve the difficult taxonomical tasks of placing countries into this contrived category. Although none are independent, many aspire to independence: India, Argentina, Pakistan, South Africa, Brazil, South Korea, and Taiwan. The latter three, however, seek that goal through intimate cooperation with suppliers across the whole range of nuclear reactor and fuel cycle activities. These governments willingly submit their programs to extensive international safeguards. South Korea and Taiwan yielded to American pressures to cancel reprocessing efforts. Although the Brazilians have resisted similar pressures, there seems to be no rush to build the reprocessing plants they have ordered from West Germany. Brazil's refusal to sign the NPT distinguishes it from the other dependents, but this must be considered in light of the Argentine refusal to do the same. Meanwhile, Brazil has gone further than its major South American nuclear rival by ratifying the Latin American Nuclear Weapons Free Zone agreement, the Treaty of Tlatelolco.

The more determined independence seekers are less pliant. The governments of Argentina, India, Israel, Pakistan, and South Africa have long clothed their nuclear endeavors in the rhetoric of national independence. The Israelis are still planning their first nuclear power project and so fall naturally into the nonnuclear category as defined in this book, as do the South Africans, who are just completing construction of their first nuclear power units, Koeberg-1 and -2. The justification for grouping them with the other independents derives from their independent, secretive, and apparently ambitious development of nuclear technology, especially as it relates to the fuel cycle. Also, both of these states are pariahs, diplomatically and atomically. This status ineluctably leaves them much more to their own devices than the nations grouped under either the nonnuclear or the dependent nuclear headings.

The countries in the 'independent' category often generate intense press interest and speculation over their ultimate atomic aims. Their prominence stems, first, from the improved capabilities, achieved in the quest for independence, to build nuclear weapons or to help others to do so. Concern arises not only from the nuclear weapons potential in these three countries, but also from the prospect of Pakistan helping Arab nations create an Islamic bomb. Many were alarmed when India signed nuclear cooperation agreements with Argentina and Libya, even though neither pact resulted in serious technological exchange. Second, these nuclear programs are prominent because their governments seek prominence. Amidst strident exclamations of sovereignty, all of them (except Israel) berate the hegemonic powers for constraining supplies of nuclear technology in order to thwart less privileged nations.

These nations, however, have far more in common than notoriety. Their approaches to the development of atomic power display striking similarities. First, Argentina, India, and Pakistan were the earliest developing countries to enter into nuclear power generation. The Tarapur power station (TAPS) near Bombay commenced operations in 1969, followed in 1972 by the Karachi Nuclear Power Plant (KANUPP) and in 1974 by the Atucha-1 unit near Buenos Aires. Planning for all three of these units began in the early 1960s. India and Argentina had initiated nuclear research efforts by 1950, inaugurating their first research reactors in 1956 and 1958, respectively. Pakistani research began later, with the 1955 creation of the Pakistan Atomic Energy Commission and 1965 commissioning of the PARR research reactor at Pinstech.

Second, the independent atomic energy efforts have all enjoyed high government priority. Given the possible utility of nuclear weapons in defense of Israeli national survival, it would be surprising if the government there did not at least explore the option. The cloak of secrecy draped over the Dimona research facilities suggests the importance the Israeli government attaches to, or at least to others' perceptions of, its nuclear program. Official statements have hewn a carefully ambiguous line. In 1968, for example, Prime Minister Levi Eshkol was quoted as saying that Israel had the technical knowledge to produce a nuclear weapon but that a significant gap existed between the technological knowledge and a weapon design.[2] In South Africa, Prime Ministers H. F. Verwoerd, B. J. Vorster, and P. W. Botha have all strongly supported nuclear development. At the 1965 inauguration of South Africa's first research reactor, Safari-1, Verwoerd proclaimed 'the duty of South Africa not only to consider the military uses of the [fissile] material but also to do all in its power to direct its uses to peaceful purposes'. Five years later, Vorster announced that the Atomic Energy Board (AEB) had developed a new process, 'unique in its concept', to enrich natural uranium, an achievement 'unequaled in the history of our country'.[3] This pattern of prime ministerial support was mirrored in Pakistan, especially under Zulfiqar Ali Bhutto and Zia ul-Haq.

Dr Homi J. Babha, father of the Indian nuclear program, enjoyed the intimate support of Prime Minister Jawaharlal Nehru. This enabled him to transform the Department of Atomic Energy into what became widely known as a 'state within a state', due to its great clout. Admiral Oscar A. Quihillalt directed the Argentine nuclear program almost continuously between President Juan Peron's 1955 ouster and 1973 triumphant return. During that time, he survived eight governments and presided over the construction of the first research and power reactors in Latin America. After the 1976 military coup, Admiral Carlos Castro Madero took up Quihillalt's mantle, and has continued to obtain governmental approval for investment in nuclear power station construction. Indeed, in countries so strained for resources, unflagging leadership support has been essential to the provision of budgets adequate to pursue vigorous atomic energy efforts.

Third, these countries share a common strategy: vociferous assertion of independence. Government leaders have consistently excoriated the nuclear weapon states for their invidious efforts to 'disarm the unarmed', relegating countries without nuclear weapons to second-class status, while continuing to augment their own arsenals of mass destruction. None has signed the Non-Proliferation Treaty, on the grounds that it perpetuates this pernicious discrimination (Argentina, India, Pakistan), risks exposure of commercial secrets (South Africa), or does not guarantee that nuclear weapons will not be acquired by hostile states (Israel). This strategy has an important corollary: use foreign help where necessary but domestic talent where possible. These governments all devoted years to the training of hundreds of technicians in European and American universities and laboratories. India and Pakistan benefited from the educational base laid down by the British Raj. Argentina's population, more than 90 percent of which is European by birth or descent, is also well-educated. Indigenous industrial or personnel capabilities could therefore be cultivated with some success.

These programs are notable for their relatively slow pace. More than twenty years passed before India and Argentina, the acknowledged atomic leaders of the developing world, inaugurated their first power reactors. In Pakistan, KANUPP was built more quickly, though commencement of the second planned power reactor, a 600 MWe light water model to be built at Chasma, has been deferred repeatedly. Israel has yet to begin power plant construction, while South Africa is just completing construction of its first two. The generally modest scale of nuclear plans advanced in the independent countries in the late 1970s is shown in Table 3.4. The nuclear targets in the independent countries are modest relative not only to other countries but also to their own overall energy plans, as Table 3.5 reveals.

Table 3.4 *Operating and Projected*[a] *Nuclear Power Capacities (in MWe)*

	Existing (1978)	Projected
Argentina	360	6,065
Brazil	0	30,986
India	587	7,174
Iran[b]	0	41,530
Israel	0	3,600
Mexico	0	14,490
Pakistan	125	8,825
Philippines	0	3,052
South Africa	0	9,844
South Korea	0	19,469
Taiwan	0	14,122
Yugoslavia	0	10,032

[a]To the year 2000. These projections have since been modified.
[b]Program now suspended.
Source: B. Fox, J. J. Stobbs, D. H. Collier, and J. S. Hobbs, *International Data Collection and Analysis* (Atlanta: Nuclear Assurance Corporation, April 1979).

It may seem paradoxical that the developing countries categorized here as most independent should have such relatively small nuclear power targets. In fact, the very quest for independence, vital to their political advocacy of nuclear development, leads atomic energy commissioners to avoid commitment to high targets for installed nuclear generating capacity. Large nuclear power programs cannot be carried out without extensive dependence. The large investments at stake intensify the vulnerability caused by that dependence. Instead, the independents concentrate on achieving fuel cycle self-sufficiency, in order to insulate themselves from foreign pressures. Even here, foreign contractors must be hired to build some facilities, such as heavy water plants for India and Argentina and a reprocessing plant for Pakistan. These arrangements have often generated controversy, but this can be advantageous to governments seeking attention. Sometimes these arrangements collapse. When Pakistan refused to accept French-proposed design modifications to neutralize the proliferation risks of plutonium separation, the French refused to go through with the reprocessing plant export.[4] In such a case, less harm is done than when a power station contract falls through, because electricity supplies do not suffer directly, and because a fuel cycle facility may cost an order of magnitude or two less than a power station.

The operating power reactors in the independent countries (as well as Israel's Dimona research reactor) are all of the heavy water variety, which do not require uranium enrichment services, the most difficult

Table 3.5 Projected Share (in 1978) of Nuclear Power in Overall Energy Consumption, Year 2000, Selected Countries (in 10^{12} Btu)

	Coal	Oil	Natural gas	Hydro	Other renewables	Nuclear	Total	Percent nuclear
Argentina	354	2,516	751	486	—[a]	398	4,505	8.8
Brazil	1,260	4,729	40	3,092	—	3,078	12,199	25
India	6,684	2,865	106	637	—	471	10,763	4.4
Iran	128	2,415	1,206	85	—	426	4,260	10
Israel	—	292	6	—	—	243	541	45
Mexico	534	2,830	1,385	554	28	1,593	6,924	23
Pakistan	232	510	764	254	—	556	2,316	24
Philippines	157	837	—	126	160[b]	266	1,546	17
South Africa	2,432	1,663	128	97	—	632	2,763	23
South Korea	945	4,055	0	56	69[c]	514	4,868	21
Taiwan	263	3,316	195	70	2[b]	1,022	6,924	23
Yugoslavia	1,596	1,020	132	532	—	800	4,080	20

[a] A dash means figure unavailable.
[b] Geothermal power.
[c] Tidal power.
Source: J. B. Fox, J. J. Stobbs, D. H. Collier, and J. S. Hobbs, *International Data Collection and Analysis* (Atlanta: Nuclear Assurance Corporation, April 1979).

step in the nuclear fuel cycle. All but Israel possess uranium mining and milling facilities, and since the early 1950s the Israelis have been extracting uranium from indigenous phosphates. India also has plants for the production of zircalloy, fuel element fabrication, heavy water production, and plutonium reprocessing. The Argentine National Atomic Energy Commission (Comision Nacional d'Energía Atómica, or CNEA) has prided itself on manufacturing its own fuel elements for its five research reactors, but has yet to complete long-planned commercial-scale facilities for zircalloy production and fuel element fabrication. It briefly operated an unsafeguarded reprocessing plant and is now building a larger version at Ezeiza.

One should not be blinded by the rhetorical flourish of these 'independent' atomic efforts. Their boasts conceal a great deal of inefficiency and delay, while their impressive appearances depend on the obfuscation of many failures. In the Indian case, every power reactor apart from Tarapur (constructed by US General Electric on a turnkey basis) has been delayed four to six years, as have the Tarapur and Kalpakkam reprocessing plants. Heavy water plant construction schedules have slipped two to four years. Meanwhile, costs on most projects have, in general, doubled. Reactors have been in operation far less than expected, with a resultant escalation in the cost of generated electricity, as shown for India in Table 3.6.

Argentine efforts have also suffered. The cost of the Atucha-1 and Embalse stations roughly doubled and quadrupled between contract signature and completion (though Argentina did not pay any of the Atucha-1 overrun). An overconfident decision in the late 1960s, to decline a German offer to build a fuel element fabrication plant for Atucha, in favor of indigenous development, left the CNEA ten years later still without a commercial fuel fabrication facility. Plans for commercial-scale heavy water production also receded continually, until agreement with a Swiss firm to build a 150 ton plant finally transformed plans into action. This is not to belittle Indian and Argentine achievements; even the most advanced nations stumbled on the road to nuclear independence. Nevertheless, it is important to distinguish propaganda from fact when the rhetoric of nuclear independence underpins governmental support for the program.

Pakistan's quest for nuclear independence began later and was catalyzed by the Indian nuclear test. Subsequently, the country's independence has become increasingly compulsory, as its nuclear program has been internationally ostracized due to the program's evidently military objective. This image arose largely from Prime Minister Bhutto's famous statement that his countrymen would eat grass in order to match Indian capabilities, and his directive that Pakistani scientists fulfill that pledge. His determination has been

Table 3.6   *Estimation versus Realization Indian Nuclear Program*

A   Estimated Actual Capital Outlay on Power Reactors (in millions of rupees)

	1964	1968	1977
TAPS	485[a]		971.2
RAPS I	340[a]	525[a]	733.4
RAPS II		581.6[a]	942.6
MAPS I and II			2,034.8[a]
NAPS I and II			2,098.9[a]

B   Expenditure Estimate for the Decade 1970–80 (in millions of rupees)

	1967	1970	1977
Heavy water plants		950[b]	1,728[c]
Nuclear fuel complex	100	130	733.8
Thermal reactors (1,000 MWe)		1,300	2,774.6
Fast breeder test reactor and Reactor Research Center, Kalpakkam		500	534.2[d]

C   Price of Output

	1964 (est.)	1977
TAPS	3p/kWh[e]	13.38p/kWh
RAPS I	2.64p/kWh	18.21p/kWh

[a]Estimated figure.

[b]For plants with a production of 400 tons of heavy water annually.

[c]For plants with an annual production of 300 tons of heavy water.

[d]Expenditure for the years 1975–78 only.

[e]p = paise; 100 paise = 1 rupee.

Sources: Indian Department of Atomic Energy and Atomic Energy Commission reports, cited in R. Tomar, 'The Indian nuclear power programme: myths and mirages', *Asian Survey*, vol. 20, no. 5 (May 1980), p. 523.

shared by his successor, General Zia. Growing alarm in the West over Pakistani nuclear intentions prompted termination not only of French assistance in reprocessing, but also of Canadian nuclear and American military and economic assistance. Bereft of spare parts and short of fuel, the KANUPP station was forced to operate below capacity, when it was in working condition at all. Some estimates indicate that KANUPP operated at a miserable load factor of 15.6 percent of capacity from 1976 to 1980, and only 5.5 percent in the latter half of that period. At this rate, KANUPP displaces less than 2 per cent of Pakistan's oil consumption.[5] A contract negotiated between the PAEC and Westinghouse of Canada for a $3 million fuel fabrication plant went by the board. In response, Pakistan built its own fuel fabrication facility, and in 1980 PAEC Chairman Munir Ahmad Khan proudly announced that Pakistan had become self-sufficient in natural uranium fuel production and would thereby save around $40 million annually in foreign exchange.[6] Unlike the Argentinians and Indians, who excluded enriched uranium reactors from their nuclear plans, the Pakistanis have retained an interest in light water reactors, and thus a justification for their most widely criticized nuclear construction project: a uranium enrichment plant at Kahota.

In early 1979, while dust from the reprocessing imbroglio was settling, surreptitious Pakistani efforts to buy the components for a centrifugal enrichment facility came to light. Dummy companies were set up to buy the necessary parts in the United States, Canada, Great Britain, and Switzerland, sometimes on the pretext that these sophisticated electrical converters and the like were required for textile production. Directing the project is Dr A. Q. Khan, who apparently had violated security regulations while working on the Urenco centrifuge project in the Netherlands, by making off with the blueprints for the plant. The Pakistanis have rebuffed foreign pressures to stop this project, insisting that their intentions are benign. The Reagan Administration claimed to have received informal assurances that, in exchange for US military assistance, Pakistan will not test nuclear weapons. Few, least of all their subcontinental neighbors, are assuaged. Dr K. Subrahmanyam, Director of the Indian Institute for Defence Studies and Analyses, noted with concern that the centrifuge project was 'a special project under the Pakistan Ministry of Defence, headed by Major-General Anis Syed', modeled somewhat on the American Manhattan Project.[7]

The South Africans have driven toward nuclear independence through uranium enrichment, a process first promoted to add value to their natural uranium export. Following the 1970 announcement of their new enrichment technology, similar to the Becker jet nozzle process, a large, five million SWU plant was planned, whose capacity could accommodate domestic needs while leaving a large surplus for exports.[8] No outside help was expected. The combination of cost escalation, financing difficulties, reduced demand, difficulty in obtaining the necessary parts from the United States and other advanced nations due to political sensitivities, and the anticipated glut of enrichment capacity in the world market convinced the South Africans to cut back their plans drastically. This decision was reinforced by the uncertainty that sufficient raw uranium feed could be mined domestically to fill the plant's capacity. Plans for the large plant were replaced by a more modest proposal to expand an existing pilot enrichment facility at Valindaba, to the scale needed to supply the two Koeberg reactors near Capetown.[9] Indigenous ability to fuel these reactors had become essential in the light of South Africa's increasing diplomatic isolation and the possibility that it could at any time be cut off from outside enriched uranium sources. Another measure of independence was enforced by the American termination of the supply of 93 percent enriched uranium-235 for the Safari-1 research reactor. In late April 1981, the South Africans announced that they had manufactured their own 45 percent enriched uranium fuel elements to run Safari.[10] Uranium of this quality is just sufficient to fashion a crude nuclear explosive.

The prospect for attaining independence is enhanced by a fourth and final characteristic common to the independent countries: favorable resource endowments. As Table 2.4 showed, South Africa, India, and Argentina possess significant uranium reserves, enough for South Africa to export extensively and for India and Argentina to supply all their own nuclear power plants, at least to the end of the century. Israel and Pakistan are less well placed, but Israel's needs to date have been minimal and the PAEC claims that it has enough uranium to keep KANUPP in operation, which is particularly important to a nuclear pariah. The skilled labor endowments of these countries also permit nuclear development, so long as it enjoys high government priority. Meanwhile, relatively favorable natural resource endowments in other energy sectors permit India, Argentina, and South Africa the 'luxury' of the large front-end investment required for nuclear power programs. Argentina is nearly self-sufficient in oil, while India produces roughly one-third of its own. Both have extensive, unexploited hydro sources. Total coal reserves in India – proved, indicated, and inferred – have been assessed at around 81 billion tonnes. By 1977, India was the seventh leading coal producer in the world, behind the United States, Soviet Union, China, Poland, and West Germany. South Africa is tenth, behind Czechoslovakia and Australia.[11] Argentina also has some coal, but so far has not needed to resort to its use. This good fortune may shape the style of nuclear development, both by relieving pressures on domestic financial resources needed for oil imports, and by relieving nuclear power of the need to bear the brunt of future energy generation expansion. Otherwise, the independent path of modest power capacity goals combined with ambitious fuel processing goals could appear intolerably expensive.

Israel, of course, does not share this favorable position, but because of the reduced scale and energy objectives of its nuclear effort, it does not require separate treatment. Pakistan does. It possesses far fewer resources, and imports 90 percent of domestic oil requirements. Coal reserves are extensive but low in quality, mostly high in moisture, ash, and sulfur content lignites which are 'usually noncoking and susceptible to spontaneous combustion'.[12] Exploitation of hydro resources will continue but is constrained by high seasonal flow variations, the priority of irrigation over power uses, and the inaccessibility of potential dam sites. Good natural gas reserves exist in the Upper Sind. Overall, Pakistani development of nuclear power policy is driven by the paucity of alternatives rather than their surfeit, and one might reasonably expect it to have followed the dependent path described below, which is more conducive to these circumstances. Pakistan's preference for the independent route can best be understood in the light of the powerful example of its Indian neighbors, and the intense regional rivalry to which Islamabad is susceptible.

Having described the characteristics of the independent nuclear programs, one is left with a nagging question: what is the value of independence? It is expensive. Diversification of energy supplies and energy autonomy both require that resources be allocated by political rather than by competitive market forces, real or simulated. The international trade theory of comparative advantage says that even if country A can produce *all* goods more cheaply than country B, both A and B could be better off if each specializes in producing the goods it makes best and trades with the other. Even under these extreme circumstances, trade is more profitable than autarky. Ideal conditions, though, do not exist in developing countries, none of which enjoys country A's enviable position of being able to produce all goods more cheaply than the competition. The OPEC nations obviously have an important comparative advantage in oil production, but the oil importing developing countries cannot hope to fulfill domestic energy needs more cheaply on their own than with the help of others. Therefore, energy independence can hurt the economy, in three ways. First, it requires that some comparative advantage be sacrificed for the sake of diversification or autarky. Second, more expensive domestic energy resources must be developed and purchased. Third, the funds devoted to these domestic energy resources, above and beyond what would have been spent on energy imports, represent an opportunity cost of capital lost from investment in other economic growth activities.

Since it costs more to build nuclear fuel facilities than to buy fuel services from large-scale suppliers, and to acquire the indigenous capabilities over many years to help build reactors than to buy them wholly built by foreign contractors, there must be compensating advantages. The most obvious is energy security. The extra costs may be viewed as insurance premiums in a policy to cushion the damaging effects of an energy supply cutoff. Another benefit could be prestige. Atomic fission is the flower of twentieth century science, and still invests those able to exploit it with an image of competence and achievement. Especially in many developing countries, where the legitimacy of the political regime is often attacked, a competent image could be used by political leaders to rally popular support, or to build nationalism by calling upon factions to unite in order to bring a challenging endeavor to fruition.

So long as a developing country depends on an industrialized trading partner, it suffers a weakened bargaining position. After Argentina had committed itself to purchase a 600 MWe CANDU reactor from the Canadians, it was forced to renegotiate financial terms and accede to more stringent antiproliferation conditions.[13] These concessions were costly and distasteful, but to insist that the Canadians abide by the terms originally agreed could have led to the suspension of the entire project, a development even more damaging politically than renegotiation. The lesson learned: avoid or reduce technological dependence in such a politically volatile field. Indigenous capability, say, to reprocess plutonium can offer a strong counter to use against otherwise indomitable powers, which fear the growth of nuclear weapons-grade stocks in Third World countries. If dependence is inescapable, at least its harmful effects can be neutralized through the political leverage afforded by possession of sensitive technologies.

THE DEPENDENTS

Other developing countries with nuclear power plans pursue a different strategy. They are less concerned with the painstaking cultivation of an independent nuclear effort, and more concerned with the rapid completion of nuclear power stations. Of course, all nations welcome independence in any field, because it increases freedom to act in accordance with one's own perceived best interests. But in weighing the costs of seeking early nuclear self-sufficiency against its benefits, many nations find that the heavy investment required could be more productively employed elsewhere. Efforts at nuclear independence in all of these countries except Brazil and Mexico are limited by the lack of enough indigenous uranium reserves to fuel all their planned reactors. Besides, where the attraction of nuclear power is perceived in its electricity output rather than in its prestige, political, or technological spillover benefits, it makes sense to purchase nuclear reactors from foreign firms experienced in nuclear plant construction. These firms can build nuclear stations more cheaply and efficiently alone than with extensive local participation, and can provide greater confidence that the completed reactor will perform well. Shared projects imply delays, mistakes, extensive training, and ambiguous burdens of responsibility. Such drawbacks are compensated by improvement of local skills; at issue is how much is gained and at what cost. The dependent nuclear countries have all concentrated on reactor installation more than fuel cycle independence.

The dependent countries adhere to the same logic for expanding electricity supplies as the independents. Growth in the use of energy is correlated to growth in national product; thus primary energy demand forecasts in these countries range up to 10 percent per annum over the next two decades. Electrical power growth is often expected to keep pace with or even exceed the rate for primary energy. Meanwhile, the costs of fossil alternatives will continue to rise unremittingly. These countries all depend heavily upon imported energy supplies, and view nuclear power both as handmaiden to growth and escape from dependence. The use of electricity is projected to expand quickly enough to accommodate the necessarily large increments of nuclear power, the nuclear units themselves providing the cornerstones for the expanded national grids. These governments have judged the comparative environmental risks acceptable, and although placing less emphasis on indigenous technological development, they hope to benefit gradually from the assimilation of nuclear and related technologies.

Of course, apart from the affinity of their energy policies, the dependent nuclear countries are a diverse group, spanning three continents and more races. *Per capita* national incomes range from $510 in the Philippines to $1,160 in South Korea. Technological abilities, industrial infrastructures, and skilled labor availability are similarly varied, as are fossil fuel resource endowments. Here again, Taiwan and South Korea pose a serious proliferation threat since, in addition to their technical sophistication, the strong threats to national survival which they face could induce these governments to acquire nuclear weapons. Despite their heterogeneity, however, it is important to distinguish these countries as a group from the independents, along the lines of program size, energy alternatives, truculence, and technology transfer techniques.

The dependent countries generally plan to introduce more nuclear power more quickly than do the independents. This was apparent in Table 3.4. Some of these plans, it is true, have fallen prey to attack, mishap, or reduction, but what is surprising is the resilience of these high targets in the face of enormous problems. Government policy vacillation plagued construction of the first two Mexican power reactors at Laguna Verde, where construction began in 1972 but had fallen six years behind schedule by 1980, 'if indeed it can be considered to have a schedule any more'. Nevertheless, the 1981 energy program confirmed Mexican optimistic intentions to build twenty nuclear plants in as many years.[14]

In Brazil, the site of the first (American Westinghouse) reactor – Itaorna, an Indian word meaning 'strange rock' – was accepted for the site of the second (German KWU) reactor without resurveying, despite serious drainage and stability problems encountered in Angra-1 construction. Instead of the shallow, solid foundations expected, the bedrock turned out to be thirty meters below surface and interrupted by Brazil's only seismic fault, a forty meter sheer drop to the continuing bedrock. As a result, the project has been delayed three years while extra foundation pilings have been sunk. The project will cost at least 50 percent more than expected. After five years of criticism at home and abroad of the delays and cost overruns which have plagued construction of both reactors and Angra dos Reis, the Brazilian government merely shifted back the scheduled completion of the nine planned reactors from 1990 to 2000.[15]

The ambitious East Asian programs have fared better. By late 1979, one Korean and two Taiwanese reactors had entered commercial operation. In an optimistic moment, Korea nearly doubled its nuclear target to forty-four reactors to be completed by the year 2000. This revised projection has since been abandoned, but in the meantime construction continues on one Canadian HWR and five Westinghouse PWRs, in addition to the operating Ko — Ri 1, near Pusan City. By 1991, thirteen nuclear plants are expected to link 11,000 MWe capacity to the grid. In Taiwan, all six of the reactors of the first phase of the nuclear program are either in operation or under construction, and construction on the second tranche of six 900 MWe units, to be completed by 1993, has also begun.

The dependent nuclear countries' energy resource pictures differ from those in their independent counterparts. The Philippines, Brazil, South Korea, and Taiwan all possess relatively poor fossil fuel endowments, and import 85 percent or more of their oil. This is a costly process. Over all, the share of energy supplies (as measured in energy equivalents) which are imported annually reaches 80 percent for Taiwan, 95 percent for the Philippines, and 100 percent for South Korea.[16] Brazil and Mexico are the best endowed of this group. Brazil has oil shale deposits second only to the United States in scale, but their development prospects remain uncertain. Brazil also has extensive hydroelectric resources, whose development commands top priority. Mexico has oil, but wishes to maximize its value by saving it for export and petrochemical uses. Significantly, these two countries, which could be considered to have better alternatives to nuclear power than the Asian dependents, also happen to have extensive uranium deposits. (See Table 2.4.) Mexico's uranium occurs near the surface and can easily be exploited. Whenever the Brazilian and Mexican governments waver in their commitments to what has become for each an extremely exasperating nuclear program, these uranium reserves help restore their vigor.

The dependent country governments behave less truculently than their independent counterparts. All have accepted international safeguards on all existing facilities, and all but Brazil are NPT members. The Brazilian government argues, not surprisingly, like the Argentinians, that the NPT is unfairly discriminatory. Brazil has been more accommodating than Argentina, however, both in accepting IAEA safeguards and in ratifying the Treaty of Tlatelolco. (Argentina has yet to redeem its pledge from the 1977 UN Special Session on Disarmament to sign this accord.) The South Korean government yielded to heavy US pressure to cancel its contract with the French for the import of a reprocessing facility, even though it had already accepted IAEA safeguards of unprecedented stringency, covering not only the plant itself but any other plant that used the same technology for fifteen years. The Taiwanese do not merely accept international safeguards, their Atomic Energy Council *insists* on them as a precondition of sale. This compliant attitude supports the preference of the dependent countries to build up their nuclear capacity with minimal hindrance.

The technology transfer policies of the dependent nuclear countries differ from those of the independents. In general, a more gradual approach is favored. The hard way of learning by doing is deferred until a period of learning by watching and studying has been completed. Often governments in the South will simply commission a consortium from the North to build a power plant or other large project, and then 'turn the key' over to local authorities. 'Turnkey' contracts are not confined to the poorer countries, and in fact many utilities in the industrialized nations have resorted to the same practice, at times vesting more confidence in an experienced reactor manufacturer than in their own construction divisions. Between 1963 and 1972, thirteen nuclear power plants were built on a turnkey basis in the United States. The turnkey approach is faster and cheaper than the independent approach. Technology transfer still takes place, but in a more gradual process of learning by watching and then, as more and more domestic labor is substituted for foreigners, learning by doing. Countries such as Taiwan and South Korea are already well along the learning curve, and their governments plan to become self-sufficient, or nearly so, in the nuclear field by the 1990s. Brazilians share this hope.

For all that, the style of efforts to achieve independence distinguishes the dependents from the independents. The Argentinians obtained 35 percent participation in the construction of their first power reactor, although this admittedly was confined to the nonnuclear parts of the project. The Korean Electric Company (KECO) did not plan to reach the 35 percent level for local fabrication of station components until construction of its fifth and sixth units.[17] These figures are not directly comparable, since the definitions and valuations of local components vary, as does the composition of the 35 percent. The Korean figure includes local fabrication of 10 percent of the core of the reactor, the nuclear steam supply system, which the Argentinians left entirely to their German contractors. Nevertheless, the point is illustrative, especially since the distinction between the independent and dependent nuclear countries stems as much from the different ways in which each type chooses to present itself as from the different ways in which each behaves.

In the Republic of China, the national electrical utility, Taipower, has a great deal of experience in building conventional power stations. So when it came to building the first nuclear station, the equipment was purchased from US General Electric, and Taipower employees built the unit themselves, with the help of foreign consultants. One Atomic Energy Council official estimated that fewer than fifty Americans came over to work on the project. At peak manpower, around 3,000 workers could be found at each site. Still, although the Argentinians obtained a technology transfer agreement from Canada with their second reactor purchase, the Taiwanese waited until inviting tenders for their ninth and tenth reactors to seek a technology transfer package. The difference between Taiwan, South Korea, and the independents is thus less one of capability than of emphasis and rhetoric. The Koreans and Taiwanese have demonstrated keen interest in fuel cycle technologies, but have subordinated fuel cycle to power development, especially when faced with US government pressure.

One vehicle for technology transfer is the joint company. The Brazilian government explicitly considered the possibility of developing 'by its own means, reactor and fuel cycle industries, with foreign technical assistance', but decided instead to 'import technology to accelerate the process of nuclear development and absorb it in medium range'.[18] To that end, its 1975 multibillion dollar agreement with West Germany involved a number of jointly-owned subsidiaries of the German reactor manufacturer, Kraftwerk Union, and the Brazilian semipublic company, Nuclebras, as Table 3.7 details.

The Korean government pursued a similar policy. After acquiring their first reactor on a turnkey basis, the Koreans decided that their large industrial group, Hyundai, should eventually achieve the capability to design and manufacture complete nuclear reactor systems in the country. In late 1978, Westinghouse announced its intention to become a minority partner in this venture, essentially to serve as conduit for the desired technology. Within eighteen months, Hyundai had received the coveted N-stamp from the American Society of Mechanical Engineers, which signified its ability to manufacture com-

Table 3.7   *Brazilian Nuclear Companies*

Name	Role	Brazilian ownership share (%)
Nuclep	Heavy components (NSSS) manufacture	75
Nuclam	Uranium exploration and mining	51
Nuclei	Build and operate demonstration Becker jet nozzle enrichment plant	75
Nustep	Commercialization of enrichment technology (in Germany)	50
Nuclen	Power station engineering services	75
Nuclemon	Heavy minerals exploitation	100

Source: 'Nuclebras Annual Report for 1979', in *Correio Braziliense* (Brasilia), April 18, 1980, pp. 9–12.

ponents to the rigid specifications required for nuclear reactors.[19] One of the independent countries, Argentina, in 1980 followed these precedents by establishing ENACE (Empresa Nuclear Argentina Centrales Electrificado), a company held jointly with KWU for design and engineering. The Argentine government owns 75 percent of ENACE. The impetus for its creation was Argentine dissatisfaction with the poor performance of Atomic Energy of Canada Limited in fulfilling the terms of its 1973 technology transfer agreement. Of course, independent nuclear fuel cycle development activities proceeded as before.

CONCLUSION

Countries never fit neatly into categories. In this chapter, a great deal of overlap has been noted between the independent and dependent countries, and occasionally between the nuclear and nonnuclear groupings. No country represents a pure or ideal case. The blurred edges, however, should not obscure the importance of these distinctions. Many developing country governments have so far been unwilling to commit themselves to the development of nuclear power. Conversely, others have embraced the nuclear power option. Of these, some stress the political significance of nuclear technology, and are sometimes willing to seek a measure of nuclear fuel cycle independence at the price of some economic inefficiency, diplomatic hostility, and reduced nuclear cooperation with suppliers concerned with nuclear weapons proliferation. The others tend to be more concerned with introducing nuclear power as cheaply and quickly as possible, with less emphasis on independent development of self-sufficiency in the nuclear fuel cycle. These distinctions are not merely cosmetic. In Part III, it will become clear how the nonnuclear, nuclear independent, and nuclear dependent countries are differently motivated.

Another point has emerged clearly in this chapter. Some of the confusion between categories results from the contagious effect of nuclear policies. Policies are sometimes negatively motivated, by governments wishing *not* to fall behind their neighbors. The atomic contagion helps explain why, for example, Pakistan has chosen the independent route. Countering the resource scarcities, which tend to favor the dependent route, has been the powerful example of its subcontinental adversary, India. The Brazilians' zealotry in their fuel cycle commitments – they ordered the Becker jet nozzle enrichment technology, even though it had never been commercially demonstrated anywhere – can be better understood in the light of the long-standing competition with Argentina.

Which approach has been most successful? That depends upon how one defines success. India and Argentina have achieved a great deal in assimilating technology and improving industrial capabilities in the nuclear field, but it has cost them plenty of labor and capital. There have been notable failures, and after thirty years of development, nuclear power still contributes only 2 percent of their energy supplies. Perhaps their greatest returns to the nuclear investment still lie in the future. Nonnuclear countries have avoided large expenditures without foreclosing the nuclear option, but if their oil situations deteriorate or their alternative energy resources cost more to develop than expected, they may later pay for their present caution with reduced income growth.

The performance of the dependents has varied, with Taiwan and South Korea faring better than the others. The variations reflect the different levels of experience brought to bear at the initial stages of nuclear cooperation. The Philippine government, for example, appears to have been severely disadvantaged by its nuclear station contract with Westinghouse, as will be shown in Chapter 10. If nuclear weapons proliferation resumes, and fuel services from the industrialized countries again become subject to political intervention, the dependent countries' governments may well regret the faith they placed in the North.

This chapter has provided only general descriptions, without indicating how nuclear policies in fact develop. The purpose of the next three chapters is to show what nuclear programs actually look like. For purposes of illustration, one country from each of the three categories has been included: Argentina from the independents, Iran from the dependents, and Indonesia from the nonnuclear countries.

NOTES

1   The Pakistan Atomic Energy Commission received provincial government approval for a 50 MWe light water reactor to supply what was then known as East Pakistan. This proposal stemmed from cost and feasibility studies conducted by Bechtel Corporation. See *Nucleonics Week*, April 18, 1963 and June 6, 1963, p. 5.
2   Stockholm International Peace Research Institute, *SIPRI Yearbook 1972: World Armaments and Disarmament* (London: Taylor & Francis, 1932), p. 312.
3   South Africa, Parliament, House of Assembly, *Debates*, vol. 25, cols. 57/8, July 20, 1970, quoted in J. E. Spence, 'The Republic of South Africa: proliferation and the politics of "outward movement"', in Robert M. Lawrence and Joel Larus (eds), *Nuclear Proliferation: Phase II* (Lawrence: The University Press of Kansas, 1974), pp. 215–16. The South African process is similar to the jet nozzle technique developed in Germany in which the uranium hexafluoride is mixed with hydrogen or helium, passed over a curved surface at supersonic velocities, and then separated into an enriched and a depleted stream by a knife edge. The heavier uranium-238 isotope is pushed to the outside by centrifugal force, and a knife edge separates this uranium-235 depleted stream from the enriched inside stream.
4   The French proposed that instead of the Purex process, which yields weapons-grade plutonium, the Pakistanis accept 'coprocessing' technology, in which the plutonium and uranium are separated out from spent fuel together. When mixed with uranium, plutonium cannot easily be converted to weapons use.
5   G. MacKerron to *New Scientist*, March 19, 1981, p. 765.
6   For details on the original agreement, see *Nucleonics Week*, July 26, 1973, p. 9; and *Nuclear Engineering International* (August 1973), p. 600; ibid. (September 1973), p. 672; ibid. (February 1974), p. 66. For Khan's statement, see the *Uranium Information Centre Newsletter* (September 1980), p. 8.
7   *Nucleonics Week*, December 4, 1980, p. 10.
8   Enrichment capacity is measured in the separative work units, or SWU, which are relative to the mass of the uranium hexafluoride gas (UF_6) pumped through isotope separation. The more uranium-235 stripped from each kilogram of natural uranium, the more separative work is required. Thus, the tails assay (amount of uranium-235 left in the unenriched waste stream) can be reduced, thereby reducing the amount of natural uranium required, by increasing the separative work.
9   R. K. Betts, 'A diplomatic bomb? South Africa's nuclear potential', in J. A. Yager (ed.), *Nonproliferation and U.S. Foreign Policy* (Washington: The Brookings Institution, 1980), pp. 289–90.
10   *The Financial Times* (London), April 30, 1981.
11   P. D. Henderson, 'Energy resources, consumption, and supply in India', reprinted in V. Smil and W. E. Knowland, *Energy in the Developing World* (Oxford: Oxford University Press, 1980), pp. 173, 179, 283; and H. H. Landsberg *et al.*, *Energy: The Next Twenty years* (Cambridge, Mass.: Ballinger, 1979), p. 283.
12   J. B. Fox *et al.*, *International Data Collection and Analysis*, prepared for US Department of Energy under Contract No. EN-77-C-01-5072, vol. 4 (Atlanta: Nuclear Assurance Corporation, 1979), p. Pakistan-1.
13   *Nucleonics Week*, September 18, 1975, p. 6; ibid., February 5, 1976, p. 8. Its unreliability as a trading partner in this endeavor gravely damaged the bargaining position of Atomic Energy of Canada Ltd, in subsequent competition for the Atucha-2 nuclear power station contract.
14   *Latin America Weekly Report*, November 21, 1980, p. 7; ibid., January 23, 1981, p. 4. Mexican Parliamentary controversy over nuclear program discussed in *Financial Times* (London), February 17, 1978.
15   *Nucleonics Week*, January 15, 1981, p. 3.
16   Energy Committee, Ministry of Economic Affairs, *Energy Policy for the Taiwan*

Area, Republic of China, Approved by the Executive Yuan, January 11, 1979, p. 1; Ministry of Energy, Republic of Philippines, *Ten-Year Energy Program, 1980–1989* (Manila: Ministry of Energy, January 1980), p. 7; for South Korea, see US Department of Energy, *International Petroleum Annual* (Washington, D.C.: Government Printing Office, 1980), p. 15.

17 *Nucleonics Week*, May 8, 1980, p. 8.

18 I. H. Marchesi, 'Brazilian nuclear development program', mimeo., Comissão Nacional de Energia Nuclear (April 1979), pp. 1–2.

19 *Nuclear Engineering International* (December 1978), p. 12; and *Nucleonics Week*, May 8, 1980, p. 8.

4   Argentina

Just one month after atomic weapons were used at Hiroshima and Nagasaki, the Argentine military junta proclaimed that the 'exceptional importance' of uranium affected 'the general interest of the nation', and that the use of the mineral 'for industrially applicable energy' could be foreseen within a relatively brief period.[1] Under the patronage of President Juan Peron, Dr Ronald Richter, an Austrian emigré from the Nazi fusion program, set up a research laboratory. In 1949, Peron established an experimental fusion plant, secluded on an island in Lake Nahuel Huapi, near San Carlos de Bariloche in northern Patagonia. He conferred executive authority over the island to Richter, in order to free the physicist from bureaucratic harassment.[2]

Over the next two years, Peron's regime grew troubled. Labor unrest, army discontent, friction between followers of the President and followers of Señora Eva Peron, inflation, and increasing political opposition afflicted the country. A joint Congressional committee in 1951 ordered the arrest of the publisher of the respected national newspaper, *La Prensa*, creating a potentially explosive situation. Reigning circles in the country were reportedly filled with 'fear, uncertainty, tension, and conflict'.[3] Since Peron faced elections within a year, he wished to quell internal opposition quickly and salvage his precarious position.

On March 24, 1951, he summoned a press conference to announce a stunning development. Perhaps coincidentally, the revelation occurred on the eve of a Washington conference of foreign ministers of the Pan-American states. Only reporters usually assigned to the Casa Rosada, the presidential residence, were invited; foreign journalists were unwelcome. Peron read a prepared statement, which was broadcast simultaneously to the nation: 'On 16 February 1951, in the atomic energy pilot plant on Huemul island at San Carlos de Bariloche, thermonuclear [i.e. fusion] experiments were carried out under controlled conditions on a technical scale.'[4] Instead of copying known processes of fission development at great expense, he continued, Argentina had risked failure in an attempt to create a better system. Uranium had not been used.[5] What was this better system? Peron explained that reactive nuclei were injected into a zone heated to millions of degrees, triggering atoms to fuse in a chain reaction. His peroration illustrated the tenor of the announcement:[6]

> I wanted to inform the people of the Republic with the seriousness and veracity which is my custom concerning a happening which will be transcendental for their future life and, I have no doubt, for that of the world. In so doing, I hope to exhort all Argentinians to collaborate in this great project which will redound so much to the enormous benefit of our country.
>
> Each one must do his part to help in so far as he is capable to bring about the final triumph of this project, destined exclusively to the greatness of our country and the happiness of her sons.
>
> Each one of us must with his knowledge, his studies, his self-effacement, his material assets, bend his work and effort toward making this effort successful. The country will owe them in the future a greatness which today we cannot begin to imagine. So be it.

Peron turned the press conference over to Richter, who boasted, 'I control the explosion. I make it increase or diminish at my desire.'[7] That the announcement was timed to distract attention from the *La Prensa* controversy or to make a splash at the Pan-American foreign ministers conference cannot be proved. That political hyperbole masked scientific triviality cannot be gainsaid. Within hours of the announcement, leading physicists throughout the world began registering their skepticism. Enrico Fermi, father of the first self-sustaining fission reaction, labeled the claim 'rather strange', while two German Nobel Laureates, Otto Hahn and Werner Heisenberg, termed it 'fantastic'. US scientists discounted the report as, in the words of one, '95 percent propaganda'.[8] These doubts stemmed from the physical impediments to attaining temperatures of several thousand degrees without fission. In all known chemical processes, any materials containing the reaction would melt and evaporate before such temperatures could be reached. For perspective, the first *uncontrolled* fusion reaction – an American thermonuclear weapon test – did not take place for another year and a half. Thirty years later, scientists still struggle to contain fusion reactions for controlled, peaceful purposes.

Nevertheless, the Peronist press lauded the alleged achievement. Peron, of course, was irked at the insulting world reaction. He claimed that he was 'not interested in what the United States or any other country thinks . . . . They have not yet told the first truth, while I have not yet told the first lie'.[9] Richter attempted to substantiate his claim by describing that the reaction was contained in a huge 'solar kiln', and conceded that the Argentinian process might not provide proportionate results on a larger scale. His boss appeared satisfied; within a week, Richter received an honorary doctorate and a Peronist Medal, pinned on by the President.

Publicly, Peron maintained a confident air. In his State of the Nation address, he asserted that 'if the experiments now being carried out keep their present pace, the Argentine Republic will have within two years their first atomic plants producing power for the entire country's electrical network'.[10] Privately, however, he must have been prey to doubt, even from within his own scientific community. He took personal charge of all nuclear research and development. Internal pressures against Richter mounted. A National Atomic Energy Commission (Comisión Nacional de Energía Atomica, or CNEA) had been established by presidential decree in 1950, partly in order to investigate Dr Richter's experiments.[11] Finally, in November 1952, the staff on Huemul Island was purged and Richter cashiered. After the 1955 overthrow of Peron's government, President Aramburu declassified the project documents, which convinced an investigating committee unanimously to brand Richter a fraud who had squandered some $70 million.[12]

Details of the Richter affair convey both the magnitude of the gaff, which generated press interest for years, and the consequent depth of Peron's humiliation. He had wanted to appear powerful, not oafish. To show that Argentina was indeed a leader in nuclear technology, Peron ordered the CNEA to hire qualified personnel, regardless of their political creed. The timing was fortuitous, as his purges of other institutions created a large pool of highly qualified unemployed. Largely due to the Richter affair, the Commission became a haven to anti-Peronists, providing a nonpartisan legacy which set the tone for much subsequent development of nuclear energy in Argentina.

BUILDING THE FOUNDATIONS

In 1952, Army Colonel Enrique P. Gonzalez was succeeded by Navy Captain Pedro E. Iraolagoitia as CNEA president. The naval captain hired able men of any political stripe, facilitating the commencement of the first serious development of nuclear technology in Argentina. Laboratories were installed for the study of cosmic radiation, elementary particles, nuclear spectrometry, isotope separation, and heavy water. Scientific-technical groups were also established for chemistry, geology, and electronics. In 1953, the first course on nuclear reactors was held. In 1955, physics and metallurgical institutes were created.[13] Within three years, Iraolagoitia had converted the Argentine nuclear program from a laughingstock into a serious, if small, research effort.

With Peron's departure, the CNEA was restructured and given autonomy in determining its objectives and regulations.[14] Exclusive Commission control over radioactive mineral ownership and production (the latter began in 1953) was enforced.[15] The Commission's objectives were redefined: to promote and perform studies and scientific and industrial applications of nuclear reactions, with minimal hazards to the public.[16] The man selected to chart the new course was another naval captain, Dr Oscar A. Quihillalt, director of the High Temperature Experimental Plant at Bariloche (reopened after Richter's ouster). Promoted to admiral, he was designated as president not only of the CNEA, but also of its board of directors, giving him the authority to shape as well as execute policy decisions.

When the CNEA decided to acquire an American-built research reactor, Quihillalt visited New York to sign the contract. Once there, he decided that Argentine scientists would be able to construct a small unit like the one he was shown at Argonne National Laboratories.[17] So, using American plans, the components of the RA-1 research reactor were manufactured in Argentina and swiftly assembled. From this time, the Commission sought independence in the nuclear field, never importing what could be produced domestically. The RA-1 became the first operating reactor in Latin America.

Few would have expected that more than eight years would pass before another research reactor would be commissioned. Hopes for rapid nuclear development, however, were soon stifled by the new president, Arturo Frondizi, who ejected Quihillalt and his board of directors. The new Commission head, Rear Admiral Helio-Lopez, did not appoint his own board, as had his predecessor, and found it uncooperative. Moreover, as part of his austerity program, Frondizi slashed the CNEA budget from nearly $10 million to about $4.5 million in 1960, the lowest level in the Commission's history.[18] Interviewees provided a consistent view of Helio-Lopez as a competent scientist, but neither an energetic leader nor a capable administrator. Beleaguered from within and without, his uneventful tenure lasted only a year and a half, before Frondizi reinstated Quihillalt.

The CNEA was again reorganized and became a direct agent of the President, who declared its programs of 'high national interest'.[19] Since 1956, responsibility had diffused as new departments were created *ad hoc* to accommodate the organization's expanding activities. Now it was divided into five divisions, covering raw materials, technology, energy, research, and radiological protection and safety. The management team appointed in 1960 survived nearly intact for over a dozen years, giving enough continuity for Argentina's nuclear program to maintain clear superiority among the agencies of the government as well as among the nations of Latin America.

ATUCHA-1

By 1964, the CNEA was considering the construction of a nuclear power station for the Greater Buenos Aires-Littoral electrical system. In April of that year, a national commission for the coordination of large electrical works authorized a feasibility study, for which President Arturo Illia appropriated $600,000.[20] Eschewing the common practice of commissioning a foreign consulting firm to provide independent assessment of major investment projects, the CNEA decided to conduct the entire study in-house. One of the study's directors attributed this decision to the desire to 'learn by doing' and to demonstrate to others that foreign consultants were dispensable for work 'that could be done perfectly well using indigenous talent'.[21] Though unstated, a probable additional reason for keeping the study inside the CNEA was the desire for better control over the results. The Commission had faith in its aims, and would not have welcomed criticisms of them by unpredictable outsiders.

Within fourteen months, the Commission completed a study which concluded that the Greater Buenos Aires-Littoral electrical power system would be large enough to accommodate a 500 MW nuclear power station, that such a station could enter operation by 1971, and could be as efficient and safe as any conventionally powered station.[22] Further, the study concluded that Argentine industry could participate in an estimated 40 to 50 percent of the construction and operation of the station. This large element of local participation would elicit the support of businessmen and government officials who wished to stimulate the scale and quality of industrial development. Adequacy of indigenous uranium resources added to the appeal of nuclear power, as did the expected stimulation of scientific and technological activities. Calculating that, in Argentina, light water reactors would cost 22 percent less than conventional power plants, the 1965 study concluded that it was 'technically feasible, economically convenient, and financially sound to install a 500 MW nuclear power plant, to supply electricity beginning in 1971 to the Greater Buenos Aires-Littoral Area'. German, Canadian, American, British, and French firms sent bids. Initially, the CNEA had to decide whether to purchase a light or a heavy water reactor. Light water stations used enriched uranium, while the heavy water versions used natural uranium for fuel. At the time, the United States was the sole world supplier of enriched uranium. Zealously independent Argentinians looked askance upon direct dependence upon their North American neighbors.

Both the British and the US suppliers offered enriched uranium reactors. Despite their strong preference for independence, the CNEA entertained bids for these reactors for two reasons. First, light water were less expensive than heavy water reactors, and were dominating the markets of the United States, Europe, and Japan. A principal author of the Atucha feasibility study recalled that the CNEA preferred independence, but not at any cost. The second reason was tactical: since more bids stiffened the competition, better terms would be offered by firms eager to gain a foothold in a new market.

The Canadian and German offers were most attractive. The German offer was for a natural uranium reactor, with 100 percent financing, 35 percent local participation, and the shortest delivery time. Its biggest disadvantage was the lack of commercial experience with the reactor design, based only on a 50 MW prototype at Karlsruhe. The Canadian natural uranium reactor design was superior in many respects.[23] It was better engineered and tested than the German alternative, as were its fuel elements. Looking toward the future, the Canadian design held greater promise for technology transfer. The German reactor core was contained in a thick metal pressure vessel, difficult to weld and cast. The pressure vessel was replaced in the Canadian reactor by a matrix of pressurized metal tubes interspersed among the fuel rods. These pressure tubes could be manufactured by Argentine industry far sooner than could a large pressure vessel.

The bid by the electrical giant, Siemens AG, of Germany was chosen for its superior financing terms, delivery time, and local participation. The better Canadian design was outweighed by the convenience of buying from a traditional trading partner, the reliability of Siemens (which had long had a branch in Argentina), and the full support for the sale by the German government – critical factors in a country where projects often foundered in political and economic turbulence.[24] The CNEA felt that, in extremity, the Germans would be more likely to carry through than would the Canadians. (This belief was confirmed by later experience with the Canadian-supplied second power reactor.)

The DM360 million ($70 million) contract signed, construction began at Atucha, a site on the Parana River about one hundred kilometers northwest of Buenos Aires.[25] The civil engineering structures (those outside the central nuclear steam supply system) were completed by 1971, but delivery of the pressure vessel and steam generators was delayed nine months. Further delay occurred when, during the final tests before starting up the reactor, technicians discovered that fuel rods were jiggling. Design changes were implemented, and 5,000 rods replaced. Without protest, Siemens paid all additional expenses as well as the financial penalties which the contract stipulated for delays. Nearly two years overdue, the Atucha-1 power station began delivering electricity on March 17, 1974. Official figures for the cost overrun were not published, but sources indicated that it approached 100 percent, for a loss of about $70 million to Siemens.[26]

EMBALSE AND THE PERONS

In 1967, the CNEA adopted a ten-year plan, later extended to 1980. Its major goals were to complete three nuclear power stations, to install most facilities necessary for the manufacture and disposition of nuclear fuel, and to stay abreast of world progress in plutonium reprocessing and breeder reactor technology.[27] During the military regime of General Ongania (1966–70) the CNEA expanded its technical base in order to meet the third objective. No breeder activities took place, but a small reprocessing facility was designed domestically, built, and operated, yielding the first plutonium ever produced in Latin America.[28] By 1971, five research reactors had entered into operation.

Success in achieving the second objective, fuel cycle self-sufficiency, eluded the Commission. This was not for want of effort. Much boasting had accompanied the development in CNEA laboratories of a new process to reduce uranium from the gaseous form used in enrichment processes to the solid form used in reactors. But technical advance did not translate easily into commercial production. Plans to increase uranium ore mining and milling and to acquire commercial heavy water production plants were deferred repeatedly from the 1960s throughout the 1970s. Uranium ore had been mined sporadically, with production increasing to over 20,000 tons in 1958, plummeting to around 5,000 tons during the Frondizi regime, soaring to 58,000 tons in 1965, and collapsing to 13,000 tons in 1966.[29] Production of the uranium concentrate extracted from the ore paralleled the undulations of the mining effort. Fluctuations continued in the 1970s, a decade which closed with neither a commercial fuel element production nor a heavy water plant in operation.[30]

Progress toward the first goal, power plant construction, commenced in 1967 with a feasibility study for a second station, undertaken on the initiative of the Cordoba province energy utility. The CNEA was authorized to call for bids on a 600 MWe station, nearly double the size of the 317 MWe Atucha reactor, by President Alejandro Lanusse in 1972. By then, competition had intensified. During the Atucha-1 bidding, some had doubted whether Argentina would in fact complete the project. Political chaos and economic stagnation tended to wreck major long-term capital investment projects, particularly those with large foreign exchange components, such as nuclear power. By 1973, with Atucha-1 nearing completion, the Argentinians had demonstrated their earnestness. Three changes in government had not prevented the project from succeeding.

Reactor manufacturers expected that the second reactor selected would indicate the government's final choice of a reactor type, and perhaps, if the Germans won again, of a permanent partner in nuclear power development. The Commission welcomed bids for heavy and light water reactors. If a heavy water reactor were again chosen, light water reactor vendors probably would abandon all efforts in Argentina. The contest was fought fiercely. The field had narrowed by two-thirds since the Atucha round, to a half-dozen bidders. The American firm, Westinghouse, was a leading contender. Ironically, the Germans, whose aggressiveness had contributed significantly to their initial success, did not even offer an HWR of the requested 600 MWe rating, although Argentine enthusiasm for the natural uranium fuel cycle had not diminished.[31] Kraftwerk Union (KWU) tried to convince CNEA officials that they, too, should go the light water route, and attempted to get the Anglo-Dutch-German uranium enrichment consortium, Urenco, to pledge to supply an Argentine LWR.

The Argentine priority on independence could no longer justify automatic preference for heavy water reactors. HWRs entailed dependence upon the Canadians, who alone continued to build and promote them. Dependence upon foreign enriched uranium supplies would be traded for dependence upon foreign heavy water supplies. Though at the time there remained only one exporter of enriched uranium, the US monopoly was doomed by impending competition from new enrichment facilities being built by Urenco and the French enterprise, Eurodif, financed in partnership with Belgium, Iran, and Italy. The only reason why use of enriched uranium would reduce independence *vis-à-vis* dependence on heavy water would be if demand grew so quickly that it exceeded even these added supplies, allowing the United States to continue to control the market.

Little public discussion had attended the Atucha-1 decision, when ministerial councils considered CNEA recommendations confidentially. This was not so for the second reactor decision, as debate erupted over the choice between reactor types, involving the universities, utilities, newspapers, television commentators, and even the 'man in the street.'[32] High stakes magnified the controversy. Energetic lobbying reflected energetic competition, with opposing sides leaking details favorable to their advocacy. Leaks provided the grist for public discussion. Moreover, after four years of press coverage on Atucha-1, the subject of nuclear power had ceased to be esoteric. Many people now felt qualified to express judgments upon nuclear technology. To the extent that controversy divided military officers, interservice rivalry might have afforded the protection for conflicting viewpoints necessary for a flourishing public debate.

A final possible explanation for the increased public debate over the second reactor was that, objectively, this decision was more difficult than that for the Atucha-1, when an excellent offer had been made with excellent terms. The persuasiveness of 100 percent financing had been unassailable. This time, though the Canadians had offered the lowest price possible, the LWR bids still cost 10 percent less. Nevertheless, the joint armed forces command then controlling the government selected the natural uranium line, and awarded the contract to a consortium of Atomic Energy of Canada Ltd (AECL) and the Italian electrical manufacturer, Italimpianti.[33] The price tag for this CANDU (for Canadian-Deuterium-Uranium) reactor was reported to be around $250 million. The reactor would serve the Cordoba province and was sited at Embalse, on Rio Tercero Lake. A prime attraction of the Canadian offer had been an ancillary technology transfer agreement, which the CNEA considered a valuable stepping stone toward independence in nuclear power production.

Before the final contract could be signed, Juan Peron returned triumphantly from eighteen years exile and recaptured the nation's highest office. But the chaos which distressed the country throughout his tenure and that of his wife and successor, Maria Estela Martinez (Isabel) Peron, wrought the worst internal upheaval the CNEA had yet known. As usual, with the change in government, Admiral Quihillalt tendered his resignation. This time, it was accepted. Peron's earlier CNEA chief and founder of the tradition of apolitical hiring, retired Admiral Pedro Iraolagoitia, succeeded the man whom he had preceded. Any hopes that the nonpartisan tradition would survive were soon dashed. In 1953, he had followed Peron's command to assemble the best agency possible, ignoring applicants' political beliefs. Now he was asked to install loyal Peronists. Former CNEA officials estimated that about sixty top Commission officials were replaced by supporters of the new government.

The results of the politicization of the CNEA were disastrous. Earlier, Iraolagoitia had presided over the Commission when it was a small research group, with a couple of hundred employees and a handful of scientists. The Commission had since begun to operate on an industrial scale, with a staff of 4,000 or so, nearly a quarter of whom were professionals.[34] Whether the retired admiral could have effectively handled the increased scale of operations even in the best of circumstances is an open question. In the event, he was denied adequate authority even to make a serious effort, as his access to his former mentor and then to Isabelita was eliminated. CNEA decision-making collapsed. Bootless management of daily affairs was attempted through large committee meetings. Only the powei program continued to progress. Its director, Dr Jorge O. Cosentino, had survived the purges due to his Peronist leanings. These had not cost him his job during the anti-Peronist regimes because of his discretion and ability, and Quihillalt's apolitical approach. Under Cosentino's direction, Atucha-1 was inaugurated and the Embalse reactor agreement was ratified, two lonely highlights in a period otherwise barren of accomplishment.

Dispute plagued the Embalse agreement from the start. Sensitized by the use of a Canadian-supplied reactor as the source of the plutonium used in the Indian nuclear test of May 1974, the Canadian government insisted that Argentina accept additional safeguards on the Embalse plant to prevent another nasty surprise. Meanwhile, Peron's economic policies generated 200 percent annual inflation and precipitated massive currency devaluations. As a result, within eighteen months, AECL had sustained some $200 million in losses.[35] AECL declared in late 1975 that, without renegotiation of financing terms, work at the reactor site might be stopped.

Haggling over the safeguards and financing issues continued into 1976, with the CNEA finally accepting a safeguards agreement banning the reexport from Argentina without Canadian permission of any materials produced in the reactor, banning also the use of supplied technology or materials for the manufacture of any nuclear explosive (peaceful or otherwise), and extending safeguards coverage from fifteen years to the (roughly thirty-year) lifetime of the reactor.[36] The March 1976 military coup delayed for another year the final approval of a renegotiated contract which cut losses to AECL and Italimpianti to $40 million apiece. By summer, however, the loss had again doubled.[37] As the project cost approached the $1 billion mark, the Canadians again asked for renegotiation. The plant had nearly quadrupled in price.

Construction at Embalse, as elsewhere in Argentina, virtually ceased in the last months of Isabel Peron's embattled government. Scores of people left the Commission, as salaries and working conditions deteriorated. To improve matters, the CNEA submitted a bill to Congress to convert itself into a state corporation, with a fixed income source and greater operational flexibility. Reportedly, CNEA autonomy had become so compromised in this period that every move, down to hiring and firing, required executive decree. The bill languished without approval for the remaining year of Señora Peron's regime.[38]

THE VIDELA REGIME AND ATUCHA-2

The CNEA was swiftly reordered under the new military government, headed by General Jorge Rafael Videla, which came to power on March 29, 1976. Navy Captain (later Vice Admiral) Carlos Castro Madero, a doctor of nuclear physics, became CNEA president. Soon he announced that the Embalse reactor was to become operational by 1981, to be followed by a third nuclear station slated for 1985 completion, and that the most important link missing from the Argentine nuclear fuel industry, a commercial heavy water plant, was to start up by 1984.[39]

Another round of housecleaning took place, primarily to reverse the one that had taken place three years earlier. Many who had left the Commission now returned to their former positions of authority. These purges were more heavy-handed than Peron's. In 1973, some CNEA officials resigned through harassment and frustration, while others were content to have their jobs transformed into sinecures. Outright firings were rare. In 1976, firings were the norm, and allegations arose that several Peronists who had become involved with the Commission disappeared.

The first orders of business were to get on with Embalse construction and with plans for a third station. In order to reduce costs, the already prepared Atucha site was picked for the location of the next reactor, Atucha-2. During Iraolagoitia's tenure, the CNEA expressed the intention to buy another 600 MWe CANDU reactor, but subsequent aggravation over the Embalse accord reduced the appeal of future cooperation with AECL. The unhappy relations with the Canadians also softened the memory of the mistakes and delays made by the Germans at Atucha-1.[40]

Meanwhile, shortage of funds delayed the entire Atucha-2 project one year. Such delays were the bane of the Commission, fundamentally threatening its central objectives. Progress was seen to be neither inevitable nor indelible. Cosentino speculated that 'it is possible to run the risk of losing [domestic capability], should an intermediate period elapse during which there were no projects approved' to satisfy increasing electrical demand.[41] Fears that a dangerous torpor might stultify the government's commitment to nuclear power were relieved when the Secretary of Energy's 1977 electrical power installation program included the contribution of the Atucha-2 station within ten years.

Presidential decrees cemented the nuclear commitment. One in October 1977 called for the installation of more nuclear power plants and the attainment of complete self-sufficiency in the nuclear field.[42] The next year, a presidential decree commenced a new round of bidding for the Atucha-2 station. In November 1978, the Cabinet appointed an interministerial commission to evaluate both a proposed CNEA fifteen-year plan and the bids for Atucha-2. The interministerial commission approved and the president ratified the plan, which guaranteed that the CNEA would remain busy for the rest of the century.[43] It gave the CNEA a mandate to build, commission, and operate four 600 MWe heavy water reactors, to be completed by 1997, as well as to build all necessary facilities to achieve self-sufficiency in the nuclear fuel cycle by that date. That such a major plan was approved despite Argentina's relentless economic plight documents the depth of the political commitment to nuclear power.

The decree committed the nation, once and for all, to the natural uranium cycle. Though coming as no surprise, this need not have happened. The picture of the world supply of enriched uranium, which had begun improving before the Embalse decision, had become brighter still. Lowered nuclear power projections everywhere, and the disappearance from the enrichment market of one of its biggest prospective consumers (Iran), made it clear that there would be, if anything, a glut of enrichment capacity. The Commission could easily have counted on supplies from Eurodif, since Iran had renounced its 10 percent share, and perhaps Urenco. Moreover, the nuclear power market was so starved for orders by 1979 that even those LWR manufacturers which had failed twice in CNEA competitions probably would have been willing to field serious bids yet again. But building an enrichment plant in Argentina would have been nearly impossible, so long as all nuclear suppliers continued their pledges not to export facilities or technology that could lead to nuclear weapon production. (Uranium enrichment was used to provide the explosive material in the Hiroshima bomb.) This would have prevented the Argentinians from completing an LWR fuel cycle domestically, so they decided instead to obtain a heavy water production plant, and obtain freedom from *all* dependence upon foreign suppliers.

Through the summer of 1979, the interministerial commission considered the bids for both the Atucha-2 reactor and a commercial heavy water plant. Agreement had been reached between the Germans and the Canadians not to relax international safeguards requirements, but to insist that the Argentine government promise to accept safeguards on all its nuclear facilities, so that the decision would be based only on the relative technological, financial, and other merits of the offers.[44] The CNEA then decided to divide the reactor from the heavy water plant contract in order to prevent, they said, too much dependence upon a single supplier.[45] The Swiss firm, Sulzer Brothers, was awarded the heavy water contract and the Germans, dropping their insistence that Argentina pledge to submit to comprehensive safeguards, won the contract for the Atucha-2 reactor.[46]

Some charged that the Atucha-2 decision had favored KWU because the German government insisted upon fewer safeguards than did the Canadian government. First, critics noted that the KWU exceeded the AECL offer in price by over $500 million, $1.579 billion to $1.075 billion.[47] Second, the CANDU reactor remained technologically superior to the KWU design. Twenty-seven CANDU stations had been built. The KWU design, a scaled-up version of Atucha-1, had never been used. It contained the large pressure vessel which, as noted, was less conducive to technology transfer than the CANDU pressure tube design. Moreover, the KWU reactor would have to be shut down for refueling, while the CANDU could be refueled while in operation and so reduce electricity generation losses. Third, the Sulzer Brothers heavy water plant offer was also claimed to be inferior to the Canadian version. Criticisms focused on the Sulzer Brothers' past performance, particularly on one of its plants, at Baroda in India, which was largely destroyed by an explosion, and another one in France which could not be put into operation. More concern arose because the plant was to produce ten times more heavy water than the largest existing Sulzer Brothers facility.[48]

Many suspected that the CNEA had divided the package into separate reactor and heavy water contracts in order to break supplier solidarity on safeguards. Because the Argentinians would be able to extract weapons-grade plutonium from the heavy water reactor fuel elements, these efforts to evade international controls elicited concern over possible intentions to develop nuclear weapons.

CNEA President Castro Madero vigorously denounced these aspersions. With respect to price, he noted that the German and Canadian packages were different and could not be compared directly. Specifically, he noted that the two proposals offered different net power levels, different fuel fabrication costs, different efficiencies of conversion from thermal to electrical energy, and different operation and maintenance costs. He discounted superior Canadian technology against superior German contract performance, remarking on 'the excellent operating experience of Atucha-1 and the advantages of using stations from the same supplier'.[49] Siemens had stuck to their agreement, absorbing a loss which might have exceeded the initial contract price of the whole project. The firm had not demanded renegotiations to increase price and safeguards requirements after the agreement was signed. By contrast, relations with AECL and the Canadian Government had been marred by aggravation caused, in the Argentine view, by repeated Canadian reneging on commitments. In short, the decisive advantages leading to preference of the German over the Canadian bid for Atucha-1 – the convenience of buying from a traditional trading partner, the strength of Siemens, and the support of the German government – had been vindicated by experience with the Canadians.

In defense of the selection of the Sulzer Brothers bid for the heavy water station, Castro Madero stated that the Indian heavy water plant explosion did not occur in the part of the plant where the heavy water was produced and was not caused by the method of production. He reiterated that the CNEA had applied 'the basic principle of diversifying sources of supply to avoid possible monopoly situations or exclusive political domination', and added that the Sulzer bid was $100 million cheaper than AECL's. Apparently, direct cost comparisons were more important for the heavy water plant than for the reactor.

All factors considered, one cannot prove that Argentina subordinated concerns for product quality and economy to desires to remain unfettered by international safeguards. Yet even Argentine officials privately admitted that the Atucha-2 decision swung on strategic and political rather than economic grounds, while after KWU had been awarded the right to build Atucha-2, disputes arose over German insistence on more extensive safeguards than were acceptable to the CNEA.[50] Clearly, wherever the Commission is able to avoid safeguards, it does.

NOTES

1   Republica Argentina, El Poder Ejecutivo Nacional (EPEC), *Decreto No. 22.855*, September 1945.
2   J. A. Sabato, 'Energía atómica en Argentina', *Estrategia*, vol. 20 (Santiago: October–December 1968), p. 334n.
3   *New York Times*, March 22, 1951, p. 1.
4   Text reprinted in *La Nación* (Buenos Aires), March 25, 1951.
5   *New York Times*, March 25, 1951, p. 1.
6   Ibid., text translated on p. 10.
7   N. Gall, 'Atoms for Brazil, dangers for all', *Foreign Policy*, vol. 23 (Summer 1976), pp. 180–1.
8   *New York Times*, March 25, 1951.
9   *The Observer* (London), March 25, 1951.
10 *Christian Science Monitor*, May 2, 1951.
11 EPEC, *Decreto No. 10.936*, May 31, 1950.
12 *New York Times*, February 19, 1956; *Daily Telegraph* (London), September 3, 1954.
13 J. L. Alegria *et al.*, *Una breve histórica de la CNEA* (Buenos Aires: Sociedad Científica Argentina, n.d.), pp. 7–8.
14 EPEC, *Decreto No. 384*, October 6, 1955; *Decreto-Ley No. 22.498*, December 1956.
15 EPEC, *Decreto-Ley No. 22.477*, December 18, 1956; *Decreto No. 5.423*, May 23, 1957.

16   See EPEC, *Decreto No. 842*, January 24, 1958, for regulations imposed on the use of radioisotopes.

17   O. A. Quihillalt, 'Argentine experiences in its nuclear programme,' mineo., ca. 1978. The electronics and control equipment still had to be imported.

18   Republica Argentina, CNEA, *Memoria Anual 1970*, p. 78.

19   EPEC, *Decreto No. 7006*, June 10, 1960.

20   EPEC, *Decreto No. 485*, January 22, 1965; J. L. Alegria *et al.*, La contribución de la energía nuclear a la solución del problema energetico argentino', in United Nations, *Peaceful Uses of Atomic Energy: Proceedings of the Third International Conference*, vol. 1 (New York: United Nations, 1965), p. 104.

21   J. A. Sabato, 'Atomic energy development in Argentina: a case history', *World Development*, vol. 1 (August 1973), pp. 30-1.

22   CNEA, *Feasibility Study: Nuclear Power Plant for the Greater Buenos Aires-Littoral Area: Summary* (Buenos Aires: CNEA, n.d.), pp. 1.2-1.3.

23   Sabato, 'Atomic development,' op. cit., pp. 34-5.

24   See EPEC, *Decreto No. 749*, February 20, 1968.

25   This price excluded charges for the heavy water and the first fuel charge, as well as the cost of the land. See *Nuclear Engineering International* (May 1971), p. 368.

26   For details about design failures, their detection, and cost to Siemens, see *Nucleonics Week*, May 31, 1973, p. 10, and October 29, 1970, p. 7; *Nuclear Engineering International* (January–February 1971), p. 83.

27   CNEA, *CNEA* (November 1971), pp. 2ff. There were other major goals, but they did not relate to the nuclear power program.

28   V. Johnson and C. Astiz, 'Latin America', in F. C. Williams and D. A. Deese (eds), *Nuclear Nonproliferation: The Spent Fuel Problem* (New York: Pergammon Press, 1979), p. 79; E. S. Milenky, *Argentina's Foreign Policies* (Boulder, Colo.: Westview Press, 1978), p. 28.

29   CNEA, *Memoria Anual 1964*, p. 17; *Memoria 1965*, p. 20; *Memoria 1967*, p. 30.

30   Ibid. For 1970s data, see the Organization for Economic Co-operation and Development, *Uranium: Resources, Production, and Demand* (Washington, D.C.: OECD Publications Center, 1977).

31   The German decision to emphasize their LWR offers might have been caused by the delegation of Siemens' nuclear activities to the subsidiary Kraftwerk Union, created in 1969 through a merger with AEG-Telefunken. By 1973, KWU had chosen the LWR to be its main line reactor, and no longer wished to build HWRs. Light water had apparently established its superiority throughout the world market, excepting Canada.

32   Quihillalt, 'Argentine Experiences,' op. cit., p. 6.

33   *La Prensa* (Buenos Aires), March 17, 1973, p. 1.

34   CNEA, *Memoria Anual 1970*, p. 79.

35   *Nuclear Engineering International* (November 1975), p. 898.

36   *Nucleonics Week*, September 18, 1975, p. 6, and February 5, 1976, p. 8.

37   *Nucleonics Week*, March 3, 1977, p. 9; *Nuclear Engineering International* (August 1977), p. 9.

38   United States, Department of State, *Telegram Buenos Aires 01462*, March 4, 1976, p. 4.

39   United States, Department of State, *Telegram Buenos Aires 04079*, June 21, 1976, p. 1.

40   Some discussion in 1977 indicated that the next station would combine the CANDU nuclear steam supply system, with its pressure tube matrix, and KWU-contracted turbogenerators, electrical engineering, instrumentation, and the rest of the plant. The aim was to wed Canadian technique to German reliability. This proposal was abandoned as unworkable.

41   J. O. Cosentino, 'The Argentine nucleo-electric programme', mimeo., ca. 1978, p. 6.

42 EPEC, *Decreto No. 3183*, October 19, 1977, art. 13. Also see comments by Castro Madero in *La Nación*, September 16, 1979, Suppl. pp. 4–6.
43 EPEC, *Decreto No. 302*, January 29, 1979, art. 1.
44 *International Herald Tribune*, April 8, 1980, p. 4. Nonnuclear-weapon states party to the NPT have to submit all nuclear facilities to the international safeguards system of controls and inspections. Non-NPT parties, such as Argentina, have to accept safeguards only on facilities provided by NPT parties. Indigenous activities or those contributed by other nonparties are not required to be internationally safeguarded.
45 Castro Madero, 'Statement at the signing of Decreto No. 2441' (n.d.), p. 12.
46 EPEC, *Decreto No. 2441*, September 20, 1979, arts. 1 and 2.
47 *La Opinion* (Buenos Aires), September 2, 1979, p. 14.
48 *Noticias Argentinas* (Buenos Aires), October 15, 1979. See also Castro Madero rebuttal to these charges in 'Statement 2441', op. cit., pp. 21–4.
49 Castro Madero, 'Statement 2441,' op. cit., p. 13. Desire to continue with one supplier, though, could just as easily have favored AECL.
50 *La Nación*, November 9, 1979, p. 30; *La Prensa*, November 7, 1979, p. 2, and November 9, 1979, p. 2.

5   Iran

In Iran, as in developing countries the world over, interest in the use of fission was sown by the US Atoms for Peace program in the 1950s. In 1957, at the opening of an American Atoms for Peace exhibit in Tehran, the Shah announced the signing of a proposed agreement for cooperation in research on the peaceful uses of nuclear energy.[1] Initial cooperation was limited to some technical assistance and the lease of a few kilograms of enriched uranium. A year later, a nuclear training center under Central Treaty Organization (CENTO) auspices was moved from Baghdad to Tehran. From that moment, nuclear energy interested the Shah personally. In 1959, he ordered the establishment of the Tehran University Nuclear Center, which the next year purchased a 5 MWt research reactor from the United States. In his 1961 autobiography, the Shah described in detail the training activities at the CENTO Institute of Nuclear Science and the possible medical, agricultural, and industrial applications of radioisotopes.[2] The International Atomic Energy Agency, as usual, sent an expert to assist in installing the reactor, and the United States pledged the standard $350,000 Atoms for Peace grant toward the cost.

As with similar projects elsewhere, little else happened. Able technicians were hard to find. In a country richly endowed with oil, with far too little installed electrical capacity (360 MW) to accommodate even a single power reactor, and no national grid, there seemed little urgency to pursue nuclear fission studies, even less to consider nuclear powered electricity generation. Consequently, it was not until 1967 that the small research reactor, over five years behind schedule, was commissioned for use. A Van de Graaf generator, purchased in 1957, was not installed for fifteen years. Even then, the laboratory was hardly furnished with power outlets, let alone the electronics necessary to perform atomic research.[3] By the early 1970s, still no clear program of research had been adopted at any level, and equipment was barely used.

Equally dormant was a small, policy-making Atomic Energy Commission, established in the 1960s. Divorced from Tehran University, the hub of the nation's minimal nuclear activity, the Commission languished. The Shah, dissatisfied with the lack of progress, asked the

Ministry of Water and Power in 1972 to assess the feasibility of building a nuclear power station within five years.[4] Yet again, the effort stagnated. The Ministry lacked familiarity with nuclear technology, and sought assistance from the Tehran University Nuclear Center, whose staff did not know how to administer commercial scale power programs. The expected symbiosis failed to materialize, and no major proposals emerged from the study. Simply finding enough experiments to keep their laboratory in use seemed to overtax the capacities of resident scientists.

Before 1964, most electric power facilities in Iran were small diesel units owned by local private and municipal companies, or attached to industrial installations. In the late 1950s, the central government initiated a hydropower-based power development program. Still, only one quarter of the population had access to electricity by the early 1960s.[5] In 1963, a power authority was created which was replaced a year later by the Ministry for Water and Power. Ten regional electrical companies were set up to help administer the growing public power sector and coordinate the existing network of small, private stations scattered throughout the country. The government nationalized many of these stations and rapidly expanded the power system during the Third and Fourth Plans.[6]

A GRAND EFFORT

The massive price increases imposed on oil in late 1973 and early 1974 presented opportunity and challenge to the government of Iran. After the 1971 departure of British forces from the Persian Gulf, the Shah had sought to establish Iran as the bulkhead of Western interests in the region. The quadrupling of oil revenues provided him extensive means with which to try to translate that wish into reality. The challenge was to find good use for an immense income in a country too little developed to absorb it. With aspirations for rapid modernization at home and increased projection of influence abroad, the Shah decided to augment drastically the project budgets in the country's Fifth Plan for economic development. The annual military budget reached $10 billion.[7] The Shah also decided to act upon his long-held belief that oil was far too valuable in petrochemical production (and for export) to continue to burn for domestic energy consumption.[8] In his words:

> The oil we call the noble product will be depleted one day. It is a shame to burn the noble product for the production of energy to run factories and light houses. About 70,000 products can be

derived from oil. We plan to get, as soon as possible, 23,000 MWe from nuclear power stations. Added to the electricity generated by our dams, this will give us one of the highest per capita supplies in the world.[9]

Three other major justifications were adduced in support of the 'recourse to nuclear energy on a grand scale'.[10] First, nuclear power was deemed the best long-term energy alternative, since water resources were too feeble for a significant hydropower program, and neither fusion nor solar technologies would mature before the end of the century. Second, nuclear energy required heavy investments which only a few countries could commit. It was 'natural' that Iran should profit from the ability to make them. Third, long-term forecasts showed that the cost of conventionally fired plants would inevitably continue to climb, leaving nuclear power in an advantageous position.

Frustrated by the failure of his earlier encouragement of nuclear technology development, the monarch decided that only a grand effort could make nuclear power become a reality in Iran. 'Critical mass' in commitment had to be achieved. To execute a program of this scale, a new, powerful, autonomous agency was needed. In March 1974, the Shah established the Atomic Energy Organization of Iran (AEOI) under Dr Akbar Etemad, and affirmed that 'the matter is of high importance so the commission will operate under my direct supervision'. Within a week, in his New Year's speech, the Shah announced that Iran would begin using nuclear power as soon as possible.[11]

Etemad faced an intimidating task. The Shah wanted results, quickly. The question was how to begin. In Argentina and India, the most successful developing countries in the nuclear field, scientific infrastructure had been developed assiduously and patiently. Experience with research reactors led to a gradual introduction of nuclear power generation. But in both cases more than twenty years elapsed before their first power reactors went critical. Etemad could not afford such an unhurried pace when he had orders to build about *twenty* power reactors in as many years. Nor could Iran emulate the developed countries' nuclear development pattern. Though the AEOI budget was large (increasing from $30.8 million in fiscal year 1975 to over $1 billion in fiscal year 1976)[12] in no other respect could Iran approach the effort mounted in even the smallest European nations. The country lacked a large electrical grid. Skilled manpower was in chronic short supply. Iranian industry was overtaxed, and could not contribute all the pumps, valves, and reinforced concrete needed in a nuclear station, let alone the specially designed components unique to reactors.

The only alternative was to purchase reactors and the services of their constructors. Nuclear reactor vendors were eager to make sales. Even before the AEOI was established, tentative accord had been reached for the French to build five 1,000 MWe reactors for Iran. Once Etemad took office, more negotiations quickly followed. In May 1974, US Atomic Energy Commission Chairman, Dr Dixie Lee Ray, visited Tehran. The next month, provisional agreement was reached for the United States to supply two reactors and the enriched uranium fuel for them. Projected cooperation was expanded in 1975, when Secretary of State Henry Kissinger and Finance Minister Hushang Ansary signed a $15 billion trade agreement, which envisaged the purchase of eight reactors for $6.4 billion.[13]

The high hopes for American cooperation fell prey to legal complications. Section 123 of the US Atomic Energy Act required that cooperation with other countries be based on an agreement for cooperation, which set forth the bounds and conditions of bilateral nuclear relations.[14] The 1957 US–Iran agreement (due to expire in 1979) covered cooperation only for nuclear research, and could not simply be extended to apply to cooperation for nuclear power development. Negotiations on the new research and power agreement became bemired in disagreement over the right of Iran to reprocess the plutonium and other elements from the spent fuel extracted from its reactors.

A major stumbling block to agreement was removed when the Shah agreed that any reprocessing plant in his region be subject to international control.[15] Etemad went further, announcing that Iran had no immediate plans to reprocess its spent fuel. By this time, however, the election of President Jimmy Carter had caused further delays. As a candidate, Carter had stressed the importance of preventing the proliferation of nuclear weapons, and his concern led to a more restrictive US nuclear export policy. Negotiations were also delayed due to uncertainty: the US Congress was considering a bill that would further restrict the conditions under which US nuclear exports would be permitted. To avoid concluding an accord that immediately would be rendered obsolete by new US statute, negotiators waited.

Official Iranian policy on nuclear weapons could not account for the delay, for the government of Iran had consistently and unambiguously rejected nuclear weapon acquisition. As early as 1961, the Shah wrote that, 'our philosophy is well expressed by the CENTO Institute of Nuclear Science, which is devoted entirely to peaceful applications of nuclear energy'.[16] The government quickly signed and ratified the Limited Test Ban Treaty of 1963 and the Non-Proliferation Treaty of 1968. All nuclear facilities were placed under IAEA safeguards. The Shah sponsored initiatives in the United Nations to bar all use and possession of nuclear weapons in the Middle East.[17] Admittedly, there was little cost in offering such a hopeless initiative, but the gesture was positive. The government did not retreat from this position after the nuclear program began in earnest. In 1975, the Iranian delegate to the Geneva Disarmament Conference reiterated his nation's renunciation of nuclear weapons. That year, the Shah branded the prospect of Iranian nuclear weapons as 'ridiculous' in the light of existing Soviet and US arsenals.[18] Rather, the monarch's stated military objective was to accumulate enough conventional strength to defeat any attack short of one using nuclear weapons.[19] For that contingency, Iran relied upon security arrangements.

Finally, talks in Tehran between President Carter and the Shah in January 1978 resolved all major outstanding issues.[20] The new agreement would neither promise reprocessing rights for Iran nor permit a US veto over them. Rather, the United States pledged that Iran would receive 'most favored nation' status for reprocessing. In other words, the American government would not discriminate against Iran as compared to other countries in requests for permission to reprocess plutonium from enriched uranium fuel of US origin. The Americans were also locked in controversy with the Japanese and the Europeans over this issue. The Iranian government agreed to accept some safeguards, beyond those required by the IAEA, which were demanded by Washington. Still, negotiations dragged on without conclusion until the Shah's ouster, after which they became irrelevant.

GERMAN COOPERATION: BUSHEHR AND BEYOND

The AEOI initiated more fruitful partnerships. In contrast to the sluggish discussions with the United States, negotiations with the Germans were swiftly concluded. Keen to gain a preeminent position in what appeared to be the best virgin market for nuclear reactors in the world, a Kraftwerk Union team offered to match any offer from competitors. In just two months, a detailed letter of intent for the supply of two 1,240 MWe reactors was drawn up with the AEOI.

This created an awkward situation. The Iranians did not want the German agreement to upstage the earlier commitment to the French. The preliminary agreement with France for the purchase of five 1,000 MWe reactors had been ratified during a June 1974 visit to Paris by the Shah and Etemad. The $5 billion accord provided for a French supply of uranium, industrial equipment, gas pipelines, and a nuclear research center.[21] At the same time, Foreign Ministers Abbas Ali Khalatbari and Jean Sauvagnargues signed an agreement for cooperation in the peaceful uses of atomic energy.

Translation of this umbrella agreement into commercial contracts had not yet occurred. In July, Iran agreed to extend a $1 billion line of credit with Banque de France for use over three years at commercial interest rates toward reactor payments.[22] But, according to AEOI officials, the French appeared ill-prepared to negotiate contracts for reactor exports. The Organization had to deal separately with manufacturers of the nuclear steam supply system, turbogenerators, and constructors of the civil works (such as the roads and houses at the reactor site). The German approach was more effective, with KWU leading the negotiations for the entire project, even though it sub-contracted large segments of the work. When alerted to the quick results of the negotiations with KWU, the French at once revived and signed a letter of intent in order to protect their market position.

As noted, the Shah wanted nuclear power to come on stream as quickly as physically possible, so as to maximize oil diversion from domestic consumption. His goal was to multiply Iranian electricity supplies tenfold by the mid-1990s, to over 50,000 MW installed capacity. Roughly half of this was to be nuclear. Meanwhile, electricity shortages were already occasioning discontent in Tehran. Consequently, KWU took the unusual step of commencing reactor site work, in August 1975, on the basis of only a letter of intent rather than a contract. Contracts for the two reactors, under construction at the Persian Gulf port of Bushehr, were signed the following summer. The capital cost for the pair, including infrastructure, was about DM7.8 billion (about $3 billion).[23]

The package was 'super-turnkey'. A turnkey project is one wherein the contractor takes all responsibility for design, engineering, and construction of a plant, and then simply 'turns the key' over to the customer, who is responsible only for operation and maintenance. Many developing countries choose turnkey contracts for projects that involve advanced techology, high quality control standards, specially skilled workers, or which for some other reason exceed the capabilities of the customer's industries. A turnkey contract drains foreign reserves more than does a contract in which local industry plays an important part, but, on the other hand, it increases efficiency, since a single, experienced contractor administers the whole project. Overall price may be lower than when inexperienced, inefficient local firms attempt to make a significant contribution.

Super-turnkey contracts exclude local participation to an even greater extent. In turnkey projects, the infrastructure of houses, port facilities, hospitals, schools, and roads are usually built at least partly by the industries of the developing country. But at Bushehr, even these items were built by foreigners. Again, the reason was the Shah's desire for speedy execution.[24] Iranian industry, though capable in many fields, was overcommitted by the revised five-year development plan of 1974. Iranian contributions to the Bushehr project might have been delayed by the need to obtain governmental approval and licenses. To avoid possible bureaucratic entanglements, the AEOI decided to steer clear of any possible governmental interference.

In June 1977, with final agreement for the first two French reactors still pending, negotiations opened with KWU for the second tranche of German reactors. The Iranians were pleased with the progress at Bushehr. Most reactors planned or under construction in developing countries and those developed countries with serious opposition to nuclear power (virtually all outside France and the Eastern bloc) fell drastically behind schedule. Work at Bushehr was running three months ahead. By November, a 'qualified' letter of intent was signed for the supply of four more 1,200 MWe pressurized water reactors, for an estimated DM19 billion ($5 billion).[25] The 'qualified' letter amounted to a firm contract for the supply of all equipment for the plants, leaving siting, civil works, and financing arrangements to later settlement. The Federal Cabinet guaranteed a DM10.8 billion credit, to cover more than half of the total price.[26]

It was more difficult to find suitable sites than sellers. Nuclear power stations are usually built along rivers or coastlines in order to obtain ample supplies of water to cool the superheated reactor core. The problem in Iran was that nearly all potential sites – along the Caspian Sea, Persian Gulf, and Karun River – were seismically active. Reactors could not be built in the most active spots, and required twice as much concrete reinforcement in the rest. Caspian sites were especially poor, because of the immense difficulty of transporting huge reactor components over the Elburz Mountains. Sites along the Persian Gulf or Karun River also required installation of long transmission lines to Tehran, Isfahan, Tabriz, Bandar Abbas, and points between, which would have been costly. AEOI officials conceded that site constraints would have limited maximum nuclear capacity in Iran to between 12,000 and 15,000 MWe, far short of the 23,000 MWe target.

The desire to find stable sites and to avoid concentrating installed power capacity along the Gulf and the Karun (source of most hydro-power potential, too), which were far from the major consumption centers of Tehran and Tabriz, led the AEOI to select inland sites for the next German reactors. The reactors, to be built near Tehran and Isfahan, would still be the light water version, with water cooling and moderating the reactor core. The turbogenerators, however, were to be air-cooled. This approach engendered some risks, since the 600 MWe turbogenerators (two for each reactor) were twice the size of the largest for which air-cooling had ever been tried in a nuclear reactor.[27] The system was costly and reduced thermal efficiency (less electricity would be produced per unit of heat generated in the reactor core), but would at least partially compensate by the reduced need for concrete reinforcement at these seismically stabler sites.

In other ways, the agreement for these air-cooled reactors was expected to differ markedly from those under construction at Bushehr. Remarkably, for a multibillion deutschmark contract, the Bushehr reactors were paid for in cash. By late 1977, however, slackened world oil demand had reduced Iranian revenues below expectations, causing a cash flow squeeze which precluded further strictly cash arrangements. Instead, the AEOI requested, and the German Federal Cabinet guaranteed, credits to finance 80 percent of the cost. Another difference was that Iranian local participation was expected to rise from 0 percent to 20 percent. Site work began in 1978.

FRENCH COOPERATION: THE LONG ROAD TO AHWAZ

The Germans had concluded arrangements for the Bushehr reactors before fiscal strain set in, and so were able to obtain cash financing terms. By 1976, Iranians were urging the French to supply reactors in return for oil supplies, to reduce the cash transfer needs. The prospect of a barter arrangement found no favor in Paris, where the government was apprehensive of becoming too tied down through long-term contracts with a single oil supplier. The proposal ran counter to the traditional French desire to maintain maximum policy independence by spreading its commitments among many states. In October of that year, President Valery Giscard d'Estaing visited Tehran, and obtained agreement for France to build two reactors as soon as possible, and six more later on.

This agreement notwithstanding, signatures for even the first two reactors proved elusive. Negotiations deadlocked on price and the method of payment. The AEOI felt that the French were asking for too much at exorbitant credit terms, a belief which was reinforced by the French sale of two reactors to South Africa for much less than the price presented to Iran. The French justified the discrepancy by noting that Iran was a new customer and had yet to establish the high degree of reliability necessary to obtain the best credit terms. Iran rejoined that, of all countries, few had better long-term prospects for the ability to meet financial obligations. Finally, France yielded on price, lowering the tag by an estimated 30 percent for the two reactors, but held firm in opposition to payment by oil barter.

Delays continued. Under the basic agreement, Iran would pay 40 percent in cash, while France extended seven-year credits for the remainder, through a consortium of Societé Générale, Banque de l'Union Européenne, and Banque Française du Commerce Extèrieure. Finalization was prevented by contention over the interest rate for the credit and over the insurance premium charged by the Compagnie Française d'Assurance au Commerce Extèrieure (COFACE), a public company under the French Finance Ministry which covered export risks. Without COFACE insurance, the French reactor manufacturers would have been unable to line up bank credits at reasonable interest rates, if at all.

With negotiations approaching a possibly fatal impasse, Giscard dispatched a trusted lieutenant, Michel Poniatowski, to overcome the remaining obstacles in Tehran. With the help of senior Finance Ministry officials, Poniatowski succeeded in closing the deal in October 1977, after more than three years delay. Separate contracts covered the financing terms, the power plants, and the fuel supply. A fourth agreement provided for the control of spent fuel arising from plant operation, a sensitive matter because of the plutonium contained in each spent fuel rod.[28] The site selected for the reactors was Ahwaz, forty-five kilometers north of Abadan on the east bank of the Karun River. Soon, a community of 2,500 had gathered to work there.

THE PROGRAM FALTERS

Having succeeded in sealing their first reactor deal, the French could begin discussing the six additional units proposed in the October 1976 agreement between the two heads of state. Reversing policy at Iranian insistence, France agreed in 1978 to buy more Iranian crude in return for the supply of the first four of these units.[29] In the spring of that year, KWU had begun preliminary talks for a seventh and eighth reactor during a visit to Tehran by President Scheel.[30] Hopes persisted that six to eight reactors would be bought from the United States. All told, trade analysts envisaged sales of up to eight reactors apiece from these three suppliers, for a total exceeding the 23,000 MWe target set by the Shah. The Shah also suggested that Sweden and eventually the Soviet Union would become involved in the Iranian nuclear program.[31] Reports circulated that the 23,000 MWe by 1994 goal might even be revised upward.

For technical reasons (such as the shortage of good sites and skilled manpower), however, it was unreasonable to expect that the nuclear plan target would even be reached, let alone surpassed. An important loss to nuclear power advocates was the resignation of Prime Minister Amir Abbas Hoveyda, after thirteen years in office. During the 1960s, prior to the existence of any nuclear program of note, Hoveyda supported a nuclear research center as a lodestone to draw young Iranian scientists together. Throughout his tenure, Hoveyda continued to support the nuclear program, to the benefit of the AEOI during inevitable Cabinet disputes.

His successor as prime minister, Jamshid Amouzegar, had been the Oil Minister and did not share his predecessor's enthusiasm for nuclear power development. The nuclear program was particularly vulnerable on the issue of its coordination with overall energy policy. As noted, the AEOI had been isolated from bureaucratic interference by grace of the Shah, so that it could surmount the impediments which had prevented nuclear power from taking root earlier. In one sense it had succeeded: reactors were at last under construction. The Organization's isolation, though, had also permitted a serious problem to worsen; the network to transmit the electricity from the station to the centers of consumption and to distribute the electricity within those load centers was not expanding sufficiently to keep pace with the construction of new generators. Tavanir, the state enterprise responsible for building this transmission and distribution network, through 1977 had installed neither the 230,000 volt line to bring power to the Bushehr site nor the four 400,000 volt lines which would carry the stations' output.[32] Obviously, better coordination between AEOI and the Ministry of Energy (of which Tavanir was a subsidiary) was needed.

Problems for the AEOI mounted in late 1977, as Prime Minister Amouzegar sponsored a small group of energy specialists and economists which charged that the Organization was not fully assessing or reporting 'the growing costs and risks involved in a substantial Iranian commitment to nuclear power'.[33] Mr Bijan Mossavar-Rahmani, energy correspondent for *Kayhan International* and a member of this group, adduced three sets of objections to the AEOI program. First, he contended that the limited availability of uranium worldwide, the political restrictions usually attached to its sale, and the uncertainties of the 'highly politicized, highly unstable and cartelized market' for enrichment services, would force Iranians to depend upon 'a small group of highly politicized and commercially aggressive suppliers'. This argument was flawed. The reduced electricity and especially nuclear power demand forecasts in the late 1970s justified confidence that supplies would be abundant, as indicated by the falling price of uranium after several boom years. A country with Iran's nonproliferation credentials – party to the NPT, selection of the most proliferation-resistant reactor, forbearance in the acquisition of enrichment and reprocessing facilities, acceptance of IAEA safeguards on all facilities – could easily purchase uranium even from countries with the strictest nonproliferation conditions. Both contracts for the first German and French reactors included the initial fuel cores and ten reloads. Iran had also purchased uranium contracts with other suppliers.[34]

With respect to enrichment services, again, the Iranians' nonproliferation credentials combined, if necessary, with the threat to buy their own plant if they could not purchase another's services, would probably have guaranteed continuous supplies. The AEOI had signed major enrichment contracts with the United States, France, and West Germany. Moreover, the Organization had purchased a 10 percent share in the Eurodif plant, built by the French at Tricastin, and planned to take a 25 percent share in a second French enrichment consortium, Coredif.[35]

Second, Mossavar-Rahmani claimed that the electrical infrastructure of Iran could not safely and reliably incorporate the planned reactors. The electrical grid was too small for thousand megawatt-scale reactors. The problem was aggravated by the lack of any reserve capacity in the Iranian national grid. Utilities usually operated stations which were not connected to the grid, so that during unexpected generator failures, the reserve station could immediately be connected to replace the loss of supply, thereby avoiding a blackout. The lack of reserve capacity was particularly serious in Iran, where the prolonged power cuts of 1976 and 1977 in Tehran and the government's inability to stop them were extremely damaging politically to the Shah's regime.[36]

Third, Mossavar-Rahmani charged that nuclear power was grossly expensive in Iran: over $3,000 per kilowatt, easily triple the installed per kilowatt costs of nuclear power in the developed countries and of gas-fired stations in Iran. This rate was also three times greater than AEOI estimates of the costs at Bushehr and Ahwaz. Part of the problem was that because installed electrical capacity was projected to reach 10,000 MW by 1980, too little to accommodate the full load of the 2,480 MW to be generated at Bushehr, these reactors would have had to have been operated below capacity. This in turn would have increased the cost of the electricity produced.[37]

AEOI officials denied these estimates of inflated costs, claiming that per kilowatt they were even less than $1,000, excluding the costs of the houses, roads, hospitals, schools, and harbor facilities provided in the contracts. Whether this was fair accounting practice is dubious, but nuclear advocates contended that these costs were unrelated to electricity but beneficial somehow to the commonweal. Without quibbling over tallies, however, it seems clear that there was in fact serious inflation related to the Iranian nuclear projects. Alternatively, AEOI officials argued that since the per kilowatt costs of smaller plants were much higher than those for the larger plants, the latter would compensate in the long run for their short-term inefficiency. They added that Iran also benefited from buying the supplier's main-line reactor – the 1,240 MWe model – because they would be assured of the advantages from continued improvements in the system. This claim seemed too modest; with commitments to buy up to six or eight reactors, any unit size the AEOI selected would have become a main-line system *de facto*. So long as the possibility for additional orders remained, the manufacturers had good reason to perfect their products.

The nuclear critics supported greater reliance upon Iran's natural gas resources. The country possesses the second largest known gas reserves (1,700 km^3) in the world,[38] and these reserves were not being well used. Of the natural gas which surfaced in association with oil production in 1973, only 42.6 percent was utilized.[39] The rest was wasted.

There were plans to increase natural gas use, but these took a back seat to the nuclear program. True, it would be expensive either to build gas pipelines or to transmit power, generated at a gas turbine located in the gas fields, over long distances to cities clamoring for the electricity. Nevertheless, this investment within the country could have provided employment and the opportunity to learn trade skills. More-over, some or all of the expenses relative to nuclear power were countered by savings in capital costs. For two immense gas-fired power stations under construction at Rey and Neka, they were estimated to range between $300 and $500 per kilowatt, or about a third of the lowest estimates for the Bushehr stations.[40]

Other advantages attributed to gas in comparison with nuclear power included its greater siting flexibility, easier location of smaller units near consumption centers (which would reduce transmission and distribution losses), reduced dependence on foreign technology and materials, reduced pollution, swifter construction, and greatly reduced foreign exchange requirements through reductions in both capital investment and the import of fuel services.

In the event, these criticisms took their toll, as Amouzegar decided in a review of the nuclear program to emasculate the AEOI, by trans-ferring responsibility for the planning, construction, and operation of the reactors to the Energy Ministry.[41] This left the AEOI concerned primarily with safety regulation and procurement of nuclear fuel and services. The Organization was able only to ensure that its vice president and director of the power program, Dr Ahmad Sotoodehnia, would head the new bureau responsible for nuclear power in the Ministry. By the end of June, the enabling legislation had been passed by the Majlis, and a Cabinet decision implemented the change.

THE END

Despite this draconian reorganization, criticism of the nuclear power program continued unabated. In October 1978, discussions on its fate were held between Sotoodehnia, the Atomic Energy Committee (the Ministers of Energy and of Finance, and the head of the Plan and Budget Organization), and other members of government. Negotiations with the French for more reactors were suspended. The Shah faced growing condemnations of profligacy in purchasing arms and reactors, as the cash squeeze worsened with the large pay raises awarded to 600,000 civil servants. As a result of these October discussions, he decided to postpone the purchase of the four air-cooled KWU reactors,[42] but work on the four reactors already under construction was to continue. Etemad resigned when the government failed to denounce reports that he was guilty of mismanagement and embezzlement. Sotoodehnia replaced him, and stated that 'there will be some place for atomic energy, but whether it will be 23,000 MWe I cannot say'.[43]

The program disintegrated. A long strike interrupted work at the Ahwaz site in November and December. Three monthly payments on the French and German reactors were missed. The Shah's last prime minister, Mr Shahpour Bakhtiar, in January 1979 announced the annulment of the contracts for the two French reactors under construction, citing Iran's natural gas reserves, the reactors' exorbitant prices, the shortage of available funds, and the likelihood that the reactors would 'be outmoded in a dozen years' as justifications.[44] AEOI gave its remaining twenty-five expatriate staff notice, and closed its offices in England, the United States, West Germany, and France.

In March, the French reactor manufacturer, Framatome (which had been working since October without payment), closed operations in Iran and brought all 450 of its remaining employees home. KWU recalled most of its staff of 2,100 from Bushehr and laid off about 6,400 of the 7,000 workers there, blaming the lack of delivery of building supplies to the site for several months. The attitude of the new Islamic Republic of Iran was summed up by the new AEOI president, Mr Fereydun Sahabi: 'We would have to bring in help from abroad [to continue the nuclear programme] which would bind us economically and industrially to those countries.'[45] He claimed that the cost for Bushehr had nearly doubled, to $7 billion. In August, KWU formally terminated the Bushehr project, with the two reactors approximately 85 and 70 percent complete. Subsequently, the only visible AEOI activity was the demand for restitution from France and West Germany for the allegedly illegitimate contracts.[46] The prospects for nuclear power in Iran had vanished.

Ultimately, the economic case for nuclear power in Iran did not rest upon specific calculations of per kilowatt costs. Rather, it rested upon a vision of Iran, vaulting to prosperity at the van of high technology, an electrified economy that would derive continued sustenance from oil sold by the ounce (for petrochemical production) rather than by the barrel (for burning). Pragmatism was not the hallmark of this vision, and it was on practical grounds that the nuclear plan was most vulnerable. It was expensive, and executed feverishly at the expense of alternatives, such as natural gas. It was so rapidly paced that technological benefits could not easily be absorbed in other economic sectors. As the vision of the Shah's government contracted to the daily struggle for survival, pragmatism prevailed.

NOTES

1 US Department of State, 'Atoms for Peace Agreement with Iran,' *Department of State Bulletin*, vol. 36 (April 15, 1957), p. 629.
2 His Imperial Majesty Mohammed Reza Shah Pahlavi Shahanshah of Iran, *Mission for My Country* (London: Hutchinson, 1961), pp. 307–8.
3 P. Sioshani, A. S. Lodhi, and H. Payrovan, 'On the question of pure or applied research in developing countries', in *Atomic Energy Organization of Iran, Proceedings of the Conference on Transfer of Nuclear Technology, Persepolis, Shiraz, Iran*, 4 vols (Tehran: AEOI, 1977), vol. 1, p. 306.
4 *Kayhan International*, December 19, 1972, p. 2.
5 R. F. Nyrop (ed.), *Iran: A Country Study* (Washington: The American University, 1978).
6 Bank Markazi Iran, *Investor's Guidebook to Iran* (Tehran: n.p., 1969), p. 92.
7 W. H. Forbis, *Fall of the Peacock Throne* (New York: Harper & Row, 1980), p. 237. This figure was for the 1978 fiscal year.
8 Pahlavi, *Mission*, op. cit., p. 288.
9 *Kayhan International*, August 3, 1974, p. 4. Also see text of the imperial decree, reprinted in *Tehran Journal*, March 18, 1974, p. 2.
10 The following points were condensed from A. Etemad and C. Manzoor, 'Le programme electronucléaire de l'Iran', *Annales des Mines* (May–June 1978), pp. 213–14.
11 *Kayhan International*, March 24, 1974, p. 1.
12 United States, Energy Research and Development Administration, 'Iran: atomic energy programme', mimeo. (October 1976), p. 3.
13 *The Times* (London), March 5, 1975.
14 United States, *Atomic Energy Act of 1954, as Amended*, sec. 123.
15 *International Herald Tribune*, August 16, 1976; *Financial Times* (London), August 18, 1976.
16 Pahlavi, *Mission*, op. cit., p. 308.
17 *Kayhan International*, July 8, 1974. Egypt and Rumania supported the proposal.
18 Ernest W. Lefever, *Nuclear Arms in the Third World* (Washington, D.C.: The Brookings Institution, 1979), p. 52.
19 *Middle Eastern Economic Digest*, March 25, 1977.
20 *Nucleonics Week*, January 12, 1978, pp. 2–3.
21 *Kayhan International*, June 29, 1974, p. 1.
22 *Kayhan International*, July 30, 1974, p. 1.
23 *Nucleonics Week*, July 8, 1976, pp. 4–5.
24 The Shah explicitly conceded that Iran would have to pay a premium to foreigners to build nuclear stations, on the grounds that it would take too long for Iranians to

complete the task. Quoted in S. K. Karanjia, *The Mind of a Monarch* (London: Allen & Unwin, 1977), pp. 233–4.

25 *Nucleonics Week*, November 17, 1977, pp. 2–3.
26 *Financial Times* (London), December 1, 1977; *International Herald Tribune*, December 2, 1977.
27 *Financial Times* (London), November 11, 1977, p. 1.
28 *Nucleonics Week*, October 20, 1977, pp. 13–14.
29 *Nucleonics Week*, June 29, 1978, p. 4.
30 *Frankfurter Allgemeine Zeitung*, April 25, 1978.
31 Quoted in Karanjia, *The Mind of a Monarch*, op. cit., pp. 233–4.
32 *Nucleonics Week*, February 2, 1978, p. 10.
33 Bijan Mossavar-Rahmani, 'Iran's nuclear power programme revisited', *Energy Policy*, vol. 8, no. 3 (September 1980), p. 195.
34 Lefever, *Nuclear Arms*, op. cit., pp. 55–6.
35 United States, Department of Energy, *Atomic Energy Programme of Iran*, mimeo. (Washington, D.C.: n.p., 1978), p. 9.
36 F. Halliday, *Iran: Dictatorship and Development* (Harmondsworth, Middlesex: Penguin, 1979), p. 285.
37 The cost of installed electrical capacity is calculated by dividing the total reactor cost by its output. When a reactor is run below its capacity, it effectively reduces the denominator, and hence increases the cost per kilowatt.
38 J. B. Fox, J. J. Stobbs, D. H. Collier and J. S. Hobbs, *International Data Collection and Analysis* (Atlanta: Nuclear Assurance Corporation, April 1979), p. Iran-1.
39 Echo of Iran, *Almanac 1977*, p. 268.
40 Mossavar-Rahmani, 'Iran's programme revisited', op. cit., p. 199.
41 *Kayhan International*, June 29, 1978, p. 1.
42 *Energy Daily*, October 13, 1978, p. 1.
43 *Nucleonics Week*, October 26, 1978, pp. 13–14.
44 *Le Monde* (Paris), January 30, 1979.
45 *Nuclear News*, July 1979, p. 72.
46 *Wall Street Journal*, April 12, 1979.

6   Indonesia

Indonesian interest in the atomic field was prompted first by concerns about the radioactive fallout caused by US thermonuclear weapon tests in the Pacific. A State Commission of Radioactivity and Atomic Energy was established in 1954 to study their effects on the health of inhabitants of this 3,000-mile long archipelago. The Commission was headed by a radiologist, Dr Gerrit Augustinus Siwabessy, who directed the Indonesian nuclear effort for the next twenty years. His study coincided with the beginning of the US Atoms for Peace program. The Radioactivity Commission, however, was designed to study, not to promote. The decision to become more active in nuclear activities was signified in 1958 by the replacement of the Commission by a Council for Atomic Energy, which advised the Cabinet. Early in 1959, an Institute of Atomic Energy was created to serve the Council by developing and regulating atomic energy uses.

Also in 1959, an IAEA mission to Indonesia evaluated the government's plans to introduce use of radioisotopes in agricultural research at Bogor and at Gadjah Mada University in Jogjakarta, to establish a radioisotope therapy unit at a projected cancer institute in Jakarta, and to develop a national center to provide health physics services at Pasar Minggu. Some Indonesian authorities optimistically felt that 'the installation of a small power reactor in a remote eastern region of the country might be feasible in the near future'.[1] The Agency report was encouraging, but concluded that outside assistance would be needed for all projects, and that a program to develop uranium resources, thought to exist on the islands of Kalimantan (Borneo) and Sumatra, should be preceded by a strengthening of the country's scientific staff and the establishment of experimental facilities.

From the outset, the shortage of competent staff plagued Indonesian nuclear efforts more than any other factor. More serious than shortages at the doctoral level were shortages of technicians and administrators. To compensate, Indonesian students were sent to foreign universities and special training seminars in more advanced countries whenever possible. The IAEA played a leading role in providing technical assistance. Bilateral technical assistance was offered by all of Indonesia's partners in nuclear development. In 1960,

Indonesian students began training at educational establishments in the Soviet Union in subjects that included nuclear physics, radio-chemistry, and radioactive isotope applications. Three dozen received training in the United States. Technical 'exchange' agreements were also concluded with France, West Germany, and Japan (though these extended also to many nonnuclear fields).

In May 1960, the possibility of introducing nuclear power was explored seriously for the first time at a seminar on electrical energy technology held at Bandung, on Java. In July, Indonesian Prime Minister Djuanda signed a nuclear cooperation agreement in Moscow with Vice Chairman Mikoyan of the USSR Council of Ministers, pursuant to earlier discussions held in Jakarta with Khrushchev. In it, the Soviet Union agreed to supply $5 million of necessary equipment, materials, and radioactive isotopes for a one to two megawatt research reactor and for a subcritical mass assembly to be used for training purposes.[2] The subcritical assembly was installed at the Gadjah Mada Research Center, but the research reactor project was unsuccessful.

Also in the early 1960s, the US government promised the Indonesians a $350,000 Atoms for Peace grant and $141,000 from the Agency for International Development. In April 1961, President Sukarno laid the cornerstone of an atomic laboratory at Bandung. The product of American assistance, a Triga-Mark II research reactor built by General Atomic, became the nucleus of the Bandung Center and, in fact, of the entire nuclear program. For fifteen years the only operating Indonesian reactor, the Triga achieved criticality in October 1964.[3] It was upgraded in power from 250 kWt to 1,000 kWt in 1971, and was devoted to uses such as isotope production and neutron physics experiments.[4]

SUKARNO BOMB THREAT

Despite the small scope of its nuclear effort, in 1965 President Sukarno asserted that Indonesia would detonate a nuclear weapon before the end of the year. With singular irony, he announced that, 'God willing, Indonesia will make her own bomb shortly', and brazenly told a World Congress Against Atomic and Hydrogen Bombs session in Tokyo that countries opposed to imperialism should make it their business to obtain nuclear weapons 'for the sake of peace and freedom'.[5] In a program where even assembling a small research reactor took years, Sukarno's belligerent rhetoric raised more eye-brows than fears. Even if the most sinister military and scientific extremists combined to administer an aggressive regime, the country's limited technical base would still reassure potential adversaries that they had little reason to fear for Indonesian nuclear weapons. The central obstacle to an Indonesian atomic weapon was obtaining enough fissile material for the warhead. BATAN could neither extract plutonium from the Triga research reactor fuel nor enrich uranium to the 90 percent uranium-235 content needed for even a crude device. These materials were not easily obtained elsewhere.

American-Indonesian nuclear cooperation was not immune to the strains attending the overall relationship at that time. As Sukarno's nuclear weapon threats and anti-Americanism increased, extension of the nuclear cooperation agreement became jeopardized. Concerns arose in both capitals that the US Joint Committee on Atomic Energy might not renew the agreement, and that safeguards obligations would not be transferred from the United States to the IAEA. If the agreement were terminated, American officials feared that Sukarno might abrogate his obligation to return, on request, the low enriched uranium supplied by the United States. If he did, the credibility of the Atoms for Peace program and the international safeguards system would have suffered 'both with those who have never sympathized with either and those who represent a new line of thought in Washington that the Atoms for Peace idea is aiding proliferation'.[6] American efforts to persuade Sukarno to accept IAEA safeguards included the withholding 'in protective custody' of the $350,000 Atoms for Peace grant promised the Indonesians.

Following the army takeover which began on September 30, 1965, Sukarno ceased to be in a position to make military threats. In October, the government accepted IAEA safeguards in principle and then received its $350,000 grant. The final IAEA safeguards agreement was signed in June 1967. Meanwhile, the Soviet-sponsored research reactor project at Serpong also suffered. It had not been started until 1965, as the Soviets had become annoyed at Sukarno's tilt toward China. They had sent the parts to the reactor, but the Indonesians did not construct a building to house the unit, so it sat unassembled in crates for years after the 1967 departure of the few Soviet technicians at Serpong. Under the new regime of General Suharto, communist assistance was no longer welcome, as it had been under Sukarno. The project was officially abandoned in 1971. It is unclear whether stinginess in government support or mismanagement by the project directors was most responsible for this failure. The important point is that, throughout the 1960s, a research reactor could not even be assembled and housed in Indonesia.

BATAN AND THE SEARCH FOR URANIUM

In 1965, 'in recognition of the growing importance of atomic energy, and in step with the progress and expansion in this field', the Institute of Atomic Energy was upgraded to the level of government department and renamed the National Atomic Energy Agency (or BATAN, for Badan Tenaga Atom Nasional).[7] Siwabessy remained director general and was promoted to Cabinet ministerial rank. He was succeeded in 1973 by Dr Achmad Baiquni, a leading Indonesian nuclear physicist since the mid-1950s and director of the atomic laboratory at Gadjah Mada University.

BATAN was given regulatory as well as promotional responsibilities for nuclear energy.[8] The director general reported directly to the President of the Republic. The Council for Atomic Energy was reconstituted by Suharto in 1968, and was scheduled to review BATAN policy quarterly. In reality, since the Council members were the most important ministers, it seldom met and mattered little. BATAN administered the Bandung Reactor Center; the Pasa Jumat Research Center, established in 1966 near Jakarta; and the Gadjah Mada Research Center, located at Gadjah Mada University in Jogjakarta, and home of the Soviet-built subcritical assembly. By the mid-1970s, about one-third of the 1,000 BATAN employees were professionals. A large proportion of their activities concerned the use of radioisotopes in agriculture, an extremely important field in a country with a burgeoning population and insufficient rice to feed it without resorting to imports.

Foreign assistance remained the keystone to Indonesian nuclear development in the Suharto era. Improved relations with the Western nations after 1965 expanded opportunities for cooperation. The clearest area of potential mutual benefit was uranium exploration. Advanced nations possessed the technical and economic resources to develop nuclear power, but sometimes lacked assured access to sufficient uranium resources. Indonesia lacked these technical and economic resources, but possessed geological formations which might have contained commercially exploitable uranium reserves. Nuclear power advocates in Indonesia recognized that in their country, so well endowed with alternative fuel resources, the availability of sufficient indigenous uranium for their program could prove indispensable in lobbying for its acceptance.

Grounds for agreement were clear: in exchange for prospecting rights, foreign partners would underwrite the costs of exploration, provide technical assistance, and support Indonesian development through employment and building the rail and telephone lines, harbor facilities, roads, and so forth, required for uranium mining and transport. Pursuant to the 1964 Atomic Energy Law, which reserved all uranium exploration and exploitation rights to the national government, the first agreements were signed with France in 1969. The French Commissariat of Atomic Energy (CEA) won exclusive rights to prospect in Kalimantan for seven years, and to recover their finds for thirty years. In 1975, the Indonesian-French team announced the discovery of ore between 0.3 and 1.0 percent in uranium content, in the Kalan area of West Kalimantan.[9] Final results, however, were disappointing. Uranium was not found in sufficient quantities and concentrations to justify its extraction and transport from remote jungle locations.

Negotiations with West Germany, whose domestic uranium reserves could barely sustain a single reactor-life, resulted in a 1976 agreement, signed in Jakarta by Foreign Ministers Adam Malik and Kurt Muller, which pledged the German Federal Office for Geology and Raw Materials to prospect with BATAN for six years in West Sumatra.[10] The Germans agreed to contribute over two million deutschmarks (less than one million dollars) in the first year and 90 percent of overall expenditures. They also promised to train Indonesian personnel. In return, the German government was guaranteed an interest in any uranium developments and in subsequent production in surplus of Indonesian requirements. This search also proved fruitless, and the German team returned home ahead of schedule. Negotiations with the Japanese for another uranium search fell through, probably because of the evident lack of success of other efforts. In January 1980, the Indonesian government shifted its hopes to the remote province of Irian Jaya, instructing BATAN to continue its uranium exploration efforts there.[11]

BUILDING AN ENERGY STRATEGY

Prior to the first five-year development plan, REPELITA I (1969–74), efforts to design an overall energy strategy were virtually nonexistent. Responsibilities in the energy sector were divided among nearly a dozen ministries, three major state companies, and two agencies (BATAN and the National Development Planning Board, or BAPPENAS [for Badan Perencanaan Pembangunan Nasional]). During the preparation and implementation of REPELITA I, BATAN and the National Electrical Authority (or PLN, for Perusahaan Umum Listrik Negara) organized seminars to discuss the possibility of constructing a nuclear power station in Java. The second seminar led to the 1972 creation, by BATAN and the Ministry of Public Works and Electric Power, of a Joint Preparatory Committee for the construction of nuclear power plants. The Joint Preparatory Committee was instructed to study and review the Java power system expansion program, nuclear power development in other countries, and comparisons of generation costs from nuclear, coal, and oil-fired plants, as well as the feasibility of geothermal power. Accordingly, the Committee directed a Seminar on the Economics and Technology of Nuclear Power in 1973, and held workshops on other aspects of nuclear power in the following years.[12]

Indonesian officials became more sensitive to the need for better overall energy policy coordination.[13] The first step was taken in research and development, with the 1973 creation of a State Ministry for Research. The first minister was the eminent Dr Sumitro Djojohadikusumo, whose ministerial service extended back to the first government in the Republic of Indonesia, in 1950. The year 1973 also witnessed the beginnings of centralized energy planning in the form of a ten-week seminar entitled 'Oil and Gas in Indonesia in the Year 2000'. The seminar was sponsored by the national oil company, Pertamina, the National Military Institute, and the Ministry of Mining's Directorate of Oil and Gas (MIGAS). At the closing ceremony, seminar chairman H. A. Gozali reported the conclusion that by the century's end the annual energy needs of Indonesia would reach four barrels per capita of oil or oil equivalent, requiring a total of 2.5 million barrels per day.[14] The final report recommended diversification in the use of energy. Specifically, it suggested that, in addition to its responsibilities for domestic fuel supply, distribution, and petrochemical industry development, Pertamina would participate 'in the management of other energy sources derived from minerals such as coal, geothermal [sic] and *uranium*'.[15] (Emphasis added.) BATAN officials were reluctant to trade their independence for a box in Pertamina's organization chart. Fortunately for them, the hope for a state oil company strengthened and consolidated into one integrated unit was not realized. The Pertamina financial crisis, which over the next three years revealed gross mismanagement of debt and over-diversification of investment, while nearly ruining Indonesia's credit standing by risking default on around $10 billion in foreign debt, did not inspire confidence in the state company's abilities to manage an even wider bailiwick.

The oil and gas seminar projections set the terms of debate for a four-day national energy seminar conducted in July 1974 by the Indonesian National Committee of the World Energy Conference. The objective of the seminar was to advise the government 'in determining national energy policies regarding energy'. About 150 participants from governmental, military, and political organizations attended the exercise, which was sponsored by the Ministries of Mining and of Public Works and Electric Power, PLN, BATAN, and Pertamina. The seminar reached the following conclusions concerning the pattern of future energy development:[16]

1 'The time had come' to determine an integrated national energy policy;
2 Given the long lead time and large investment required, the Government of Indonesia should develop a twenty-five year plan;
3 Energy sources should be diversified;
4 In 2000, per capita annual energy consumption should be four barrels of oil equivalent consumed for commercial purposes. The role of electrical energy in commercial energy consumption should increase from the present level of 15 percent to 40 to 59 percent. To the extent possible, natural gas should be used instead of oil for heating; and
5 From 1,000 MW in 1973, installed electrical generating capacity should reach 64,000 MW in 2000, to be divided as follows: 6 to 8 percent from hydroelectric sources, 20 to 61 percent from oil, 1 to 9 percent from geothermal sources, 12 to 25 percent from coal, and *23 to 39 percent from nuclear energy*. (Emphasis added.)

This time, BATAN had good reason to be pleased, although the tremendous range within each source projection rendered its significance doubtful.

A February 1975 symposium on energy, resources, and the environment was convened in order to synthesize the conclusions of earlier conferences and thereby provide the basis for actual policy decisions. President Suharto showed his interest by opening the three-day meeting. Policy recommendations would depend upon the amount of energy demand projected over the next decades, an issue that, not surprisingly, became central to the debate. The annual four barrels oil equivalent per capita projection came under fire from both directions. MIGAS Director Wijarso observed that the estimate might be too high, in the light of the enormous requirements in capital investment, domestic distribution, and manpower. Conversely, Dr A. Arismunandar, from PLN, argued that the four barrel estimate was, if anything, too conservative.

The symposium concluded that the government should vigorously pursue a nuclear power program in order to lay the foundations for the installation and operation of a nuclear power station by 1985. Less helpful to nuclear proponents were recommendations to reinvigorate coal production and to devise a plan to develop hydro, geothermal, and solar energy. Research Minister Sumitro met with President Suharto on March 17, 1975 to discuss energy development, afterwards telling newsmen that it would take ten years to build a nuclear power station. The following day, the Cabinet-level Economic Stabilization Council decided that the government of Indonesia would start intensive research and development programs on energy sources such as coal, geothermal, hydro, and nuclear. The government also merged the various government-sponsored research and development institutions under a single Agency for Research, Science, and Technology, expanding the authority of the State Minister over study which had previously been carried out in other ministries.

In June, President Suharto created a thirteen-member National Committee for Inventorying and Evaluating National Resources, to assemble geological data in order to gain, for the first time, a clear picture of all of the nation's resources. One subcommittee – the Energy Technical Committee – was directed to formulate a national energy policy. Three years later, the long-awaited streamlining of energy authority under a single head came to pass with the creation of the Ministry of Mines and Energy. The new Ministry incorporated the old Ministry of Mines and took authority for the electrical authority, PLN, away from the Ministry of Public Works and Electric Power. It also gained jurisdiction over Pertamina, through MIGAS; the state coal company, PN Batubara, through the Mines Directorate; and the Geological Directorate, which oversaw the geological survey of Indonesia. BATAN escaped the new Ministry's grasp.

THE NATIONAL ENERGY MIX

Indonesia's impressive array of natural energy resources toughened the task of the pronuclear lobby. According to an IAEA study, hydroelectric potential was 31,000 MW. Estimated reserves for coal were 1.2 billion tons; for oil, 15 billion barrels (2.9 billion tons of coal equivalent); and for gas, 220 million tons of coal equivalent. On Java alone, projected geothermal potential was 7,000 MWt, with 5,000 MWt more for Sumatra and Sulawesi.[17] Because Indonesian oil was of high quality, much was exported, while Middle Eastern oil was imported for domestic use.[18] Indonesian firms had an interest to reduce the use of oil in domestic electricity production, in order both to preserve domestic oil reserves for lucrative export and to reduce the burgeoning expense of Middle Eastern oil imports. The problem was how to increase electricity production without burning oil.

A succinct case for nuclear power development as an oil preservative and handmaiden to overall economic development was offered by BATAN officials, when they wrote that:

nuclear power would play a significant role in accelerating the growth of industry. It would reduce the share of oil to be burnt as fuel, while forming an indispensable addition to new coal generating stations which are to be constructed in the early eighties and to the already projected hydro power stations and geothermal plants. Under this development pattern, less developed regions, geographically more or less isolated from the richer island of Java, could benefit from the experiences, manpower, and revenues obtained from enterprises operating in the most populated area of the Archipelago.[19]

On the other hand, they warned, failure to add to the electricity grid two 600 MWe nuclear units in 1985 and 1986 'would correspond to a decrease of 2.9 to 4.4 percent in the growth of gross domestic product'.

The solution urged by nuclear proponents was the rapid development of all available energy sources, based on their very large forecasts of energy demand. When the unavoidable issue of direct competition among alternatives was raised, however, BATAN officials found the others lacking. Indonesian coal reserves were criticized for the large capital expenditure required to exploit them, as well as for their low heat content and remote location. Hydropower was also condemned for its inconvenient location to load centers. This seemed puzzling given the desire to relieve overcrowding in Java: could not coal or hydro reserves in a remote area serve to support a new center of consumption? Hydro and geothermal sources were also criticized as not always being sufficient to satisfy a region's needs, a feeble charge in light of the increased government priority on diversifying energy consumption, and certainly no reason not to develop these sources as far as possible.[20]

Criticisms of the remote jungle location of alternative energy sources also seemed specious, due to the apparent lack of recoverable uranium reserves, despite ten years of prospecting. Coal had to be shipped about 300 miles across the Java Sea from Kalimantan or across the straits from Sumatra, and the money invested in its extraction and transport would remain within the Indonesian economy, contributing to infrastructure and increasing business for local industry. Uranium would probably have to come from Australia, the United States, or Canada, and supply could be contingent upon disagreeable political conditions and safeguards requirements.

Strong cases were made for the expansion of coal and natural gas production. Until the Second World War, Indonesia had been a leading world coal producer. The memory of the two million tons produced in 1940 faded during the cheap oil era, as production bottomed at under 150,000 tons in 1973.[21] Coal production revived along with plans to reestablish its preeminence in the mid-1970s. As in Iran, much natural gas was wasted through flaring. Usually found as a byproduct of oil, gas was expensive to collect and transport, discouraging possible entrepreneurs from investing. After the oil crisis, though, such waste could no longer be countenanced, and large natural gas projects were planned.[22]

PLANNING THE FIRST NUCLEAR POWER STATION

In 1974, Indonesia requested that an IAEA survey be made of the prospects for nuclear power in Indonesia. The study was prepared by two members of the Joint Preparatory Committee for nuclear power, one each from BATAN and PLN, along with IAEA staff. In its own words.

> the major objective of this study was to determine the size and timing of nuclear power plants that could, on economic grounds and in combination with other thermal and hydroelectric power stations, justifiably be built in Indonesia on the island of Java during the study period 1978–1997.[23]

This straightforward statement betrays an implicit bias: it assumed the *desirability* of nuclear power for Indonesia by only dealing with the method and scale of its introduction.

The conclusions of the *Nuclear Power Planning Study* proceeded from a number of other assumptions consistent with this bias. Four percent annual inflation in the costs of reactor construction was assumed in a year when the actual rate reached 20 percent. (In 1980, the rate was still 16 percent.) External and social costs (undefined in the study) were disregarded. The frequency stability analysis, used to determine the largest feasible unit size (larger units are generally more economical) neglected the constraints imposed by Java's limited system of transmission lines. The inadequacy of the transmission network posed some of the greatest difficulties for centralized electricity generation in the archipelago (p. 87 of Study). The study forecast that peak demand through 1996, assuming the PLN-estimated load factors ranging from 0.68 to 0.61 during the study period, would climb somewhere between 2,300 percent and 4,100 percent. This phenomenal growth assumed that in 1973, 'suppressed demand' – in other words, the amount of added demand which there would have been in an adequate electrical supply system – was nearly double the level of the electrical demand actually satisfied, through consumption (pp. 43, 61). Of this suppressed demand, less than one-fifth represented waiting lists for additional power and new connections. The rest represented 'voltage suppressed by the overloaded distribution network' and restriction of industrial consumption during hours of peak demand. These last two factors presumed that the electricity generation system should have been designed to accommodate the peak demand naturally arising from its customers, rather than attempt to distribute the load evenly throughout the day by offering lower rates for off-peak consumption. This presumption might have contributed to the exaggeration of the demand forecast.

The study concluded that from eight to eighteen reactors could be constructed on Java from 1978 to 1992. This result could only be reached by ignoring such factors as availability of financing and skilled manpower, which were in extremely short supply in Indonesia. BATAN Director General Baiquni excused this omission by claiming that the purpose of the study was merely to ascertain 'whether the construction of nuclear power plants could bring economic benefits in the long run and, therefore, whether the commissioning of a feasibility study for the first nuclear power plant would be justified'.[24] In fact, the expressed objective of the *Nuclear Power Planning Study*, directly quoted above, was less modest than Baiquni admitted.

The debate over reactor types was less intense in Indonesia than Argentina. The desire to rely strictly upon indigenous uranium resources rather than upon American enrichment services, perhaps combined with the Indian and Argentine precedents in opting for heavy water technology (which is more economical at smaller scale than is light water technology), led BATAN officials to prefer heavy water reactors. A 1974 Jakarta workshop concluded that these reactors were most suitable for Indonesia because they both used natural uranium and offered the best prospects for local participation. That year, the Joint Preparatory Committee examined possible reactor locations along the Java coast. (River flows on the island were considered too limited to supply adequate cooling for reactors.) In 1979, Baiquni announced that two suitable reactor sites, Rembang and Lasem, had been selected.[25]

A feasibility study for the first nuclear station became the next order of business, and one that proved difficult to accomplish. Nuclear power, which requires large capital investment and foreign contracting, appeared a likely target when the Pertamina crisis enforced financial retrenchment. Unable to obtain funding from the Indonesian government, BATAN officials approached international financing institutions, such as the UN Development Program, the World Bank, and the Asian Development Bank, but to no avail. The reticence of these institutions to contribute in any way to an Indonesian nuclear power program was reinforced by the findings of independent studies for Java's electrical expansion program, that nuclear power could not become economically attractive until the mid-1990s.

The third five-year development plan, REPELITA III (1979–84) highlighted the decline in priority of nuclear power. It emphasized oil and natural gas exploration and development, and slated coal production to increase 150 percent during the planning period. Nuclear technology was mentioned, and then only in passing, among the medium-term objectives:

> to diversify energy consumption through efficient use of energy, utilization of hydropower, increasing the production of firewood, and utilization of the waste products of foreign industries as a source of energy. Research activities for the utilization of solar energy and nuclear energy will be encouraged.[26]

In the competition for electricity generation between coal and nuclear power, plans to build a large coal-fired power station in West Java, which would use up to 2.5 million tons annually, enjoyed more support than did plans for a nuclear power station.[27] BATAN officials must have been disheartened.

Nevertheless, BATAN persisted. In March 1978, the Italian firm, Nuclears Italiana Reacttori Avancati (NIRA), began to prepare the long-awaited reactor feasibility study, finally made possible by Italian financial assistance.[28] Baiquni told reporters that, despite the Harrisburg nuclear accident, nuclear energy remained the only answer to Indonesian economic development problems in the coming decades. The right time for a nuclear power plant, he continued, would be when all Javan electricity grids had been interconnected, around 1990. Admitting that BATAN did not have the 100 to 120 experts needed to run a 600 MWe nuclear plant, Baiquni suggested that this problem could be solved by training programs which had been initiated with several universities.[29]

At about the same time, a five-year scientific research and technical development accord was signed by Research Ministers B. J. Habibie of Indonesia (a supporter of nuclear power) and Pierre Aigrain of France. The French promised immediate direct aid in the construction of a 4 MW research reactor at a new science and technology complex at Serpong.[30] Finally, using the parts left behind by the Soviets over a decade earlier, together with parts discarded from the Triga reactor when it was upgraded in power in 1971, BATAN completed its second research reactor in 1979. By early 1981, the Indonesian government was making a final decision between bidders for a 300 MWt prototype reactor. BATAN officials hope that this intermediate scale reactor will serve as the final stepping stone to a full-scale, commercial power reactor. Despite the bureaucratic bruising nuclear power has endured, there seemed by that time to be a growing consensus within the energy policy-making community that it would eventually be necessary. If healthy oil earnings can be maintained, and coal production expansion continues to fall short of expectations, BATAN's hopes may be realized before the turn of the century.

NOTES

1 *IAEA Bulletin*, vol. 1 (July 1959), p. 6.
2 *Soviet News*, March 25, 1960 and July 6, 1960.
3 United States, Energy Research and Development Administration, 'Indonesian atomic energy program' (September 1975), p. 1.
4 IAEA, *Research Reactor Utilization* (Vienna: IAEA, 1972), pp. 33-7, 275-7.
5 *Washington Post*, August 2, 1965, pp. A14, B8.
6 *New Scientist*, April 22, 1965, pp. 221-2.
7 *Indonesia Today*, December 1965, p. 11.
8 Republic of Indonesia, BATAN, *Laporan visuil kegiaton BATAN* (Jakarta: BATAN, April 1977), p. 11.
9 *Mining Annual Review* (1975), p. 393.
10 *Nuclear News*, August 1976, p. 64; *Indonesian News* (London), November 1976, p. 18, and April 1978, p. 10.
11 Antara News Agency, 'Uranium exploration', January 15, 1980.
12 BATAN, *Seminar teknologi dan ekonomi pusat listrik tenaga nuklir* (Bandung: BATAN, 1973), and *Teknologi pusat listrik tenaga nuklir* (Jakarta: BATAN, 1975).
13 See R. F. Ichord, Jr, 'Indonesia', in G. J. Mangone (ed.), *Energy Policies of the World*, 2 vols (New York: Elsevier, 1977), vol. 2, p. 41.
14 US Department of State, *Airgram Jakarta A-61*, April 23, 1975, p. 2.
15 Ibid., enclosure A, p. 4.
16 Ibid., p. 3.
17 IAEA, *Nuclear Power Planning Study for Indonesia* (Vienna: IAEA, 1976), pp. 25, 33, 38, and 39.
18 Ichord, 'Indonesia,' op. cit., p. 87, explains why this policy was supported by the Indonesian ruling elite.
19 J. Iljas and I. Subki, 'Nuclear power prospects in an oil and coal producing country', in IAEA, *Nuclear Power and Its Fuel Cycle: Proceedings of an International Conference, Salzburg, 2-13 May 1977* (Vienna: IAEA, 1977), vol. 6, pp. 104-5.
20 For BATAN criticisms of alternative energy technologies, see Budi Sudarsono and Ir Prayoto, 'The role of nuclear power in Indonesia's power planning', in *Transactions: The Second Pacific Basin Conference on Nuclear Power Plant Construction, Operation, and Development*, vol. 29 (La Grange Park, Illinois: American Nuclear Society, 1978), pp. 12-13.
21 IAEA, *Nuclear Power Indonesia*, p. 32.
22 *Indonesian News*, February 1975, pp. 4-5.
23 IAEA, *Nuclear Power Indonesia*, p. 71.
24 A. Baiquni and B. Sudarsono, 'First steps toward a nuclear power programme: the Indonesian experience and prospects', in *Proceedings: The First Pacific Basin Conference on Nuclear Power Development and the Fuel Cycle, Honolulu, 11-14 October 1976* (La Grange Park, Illinois: American Nuclear Society, 1976), p. 49.
25 *Kompas* (Jakarta), trans. Foreign Broadcast Information Service, June 1, 1979, p. 12.

26 Republic of Indonesia, Embassy of the Republic of Indonesia, Washington, D.C., *REPELITA III: The Third Five Year Development Plan* (Washington, D.C.: n.p., 1979), pp. 39–41.

27 *Indonesian News*, July 1977, pp. 15–16.

28 *Energy Daily*, February 27, 1979; *Nuclear Engineering International*, July 1979, p. 6.

29 *Kompas*, trans. FBIS, June 1, 1979, p. 12; Antara News Agency, 'Energy official on consideration of new nuclear plant', February 19, 1980.

30 *Kompas*, trans. FBIS, May 9, 1979, p. 5.

Part Three: Reasons

7 Security Objectives

Nuclear policy is always decided at the highest levels, in the context of a government's broadest objectives. Any government's first concern is self-preservation, so the concept of national security tends to become closely identified with governmental stability. This chapter seeks to describe how military, political, and economic components can be affected, beneficially or otherwise, by energy and specifically by nuclear policies. The nuclear issue can be a powerful source of influence for otherwise weak nations. OPEC's success resulted from a high degree of cooperation among the few governments which provided most of the petroleum imported in the world outside communist areas. Even a resource-poor developing country acting alone, however, can improve its bargaining position with the industrialized nations by exploiting issues to which the latter are extremely sensitive. With the seizure of the Pueblo crew in the Sea of Japan and of US embassy personnel in Tehran, relatively weak governments dramatically turned the tables against the United States. Alienated governments need not always await such targets of opportunity. One Achilles' heel of the advanced nations, so sensitive that developing countries can almost always count on eliciting at least a strong response and at most major concessions, is nuclear technology. In this area, even such a fanciful threat as Indonesian President Sukarno's pledge to test a nuclear weapon can excite worldwide interest. Sukarno's assertion even became an issue in the 1964 US presidential campaign.

The mutability of objectives complicates analysis. Objectives evolve sometimes gradually, through careful deliberation, and at other times suddenly, through reactions to events. Domestic political events can alter them. In Argentina, the Peronist sympathizers who emerged in more powerful positions after 1973 sought to redress the abject dependence upon the North which grated against their anticolonial sensitivities. When he became president, Suharto shifted the emphasis of nuclear policy from militant confrontation to balanced economic development. When Iranian oil revenues quadrupled, the Shah's nuclear program leapt into activity. More typical were the many occasions on which nuclear planners have had to stanch budgetary pressures caused by economic tribulations.

Foreign events influence objectives. Traditional Latin American, Middle Eastern, and Asian rivalries extend to nuclear technology. This rivalry certainly reinforces objectives, whether or not it transforms them. Geographically remote events sometimes have immediate political effects, particularly in the nonproliferation field. Stung by the Indian nuclear test, governments from nations which export nuclear technology agreed, in a series of closed sessions in London, to guidelines which extended the use of safeguards to parts of the fuel cycle not previously covered (such as heavy water production, essential to the manufacture of the plutonium used in the Indian blast). During the Ford and Carter Administrations, US nonproliferation policy became more restrictive, with the embargo of any exports that could permit uranium enrichment and plutonium reprocessing, and the enactment of a law which required the United States to terminate nuclear cooperation with any nation that did not submit all of its nuclear installations to IAEA safeguards.[1] The following treatment of objectives, however, is not fundamentally concerned with their evolution. Instead, it analyzes those objectives which prevailed for the most formative periods in each program. Thus scant attention is given the early, dormant phases of many nuclear programs. Before turning to specific countries, the possible effects of nuclear power programs upon their security objectives is summarized.

MILITARY ASPECTS

Nuclear policies affect many dimensions of security. *First*, a nuclear power program can be used militarily to help develop nuclear weapons (or convey the impression that a government is doing so) because the rate of atomic fission can be regulated to generate either steam in a power plant or explosion in a weapon. Civil cannot be perfectly isolated from military nuclear technology. The same process used to raise the percentage of the fissile (easily split) uranium-235 isotope from its natural 0.7 percent concentration to the 3.0 percent or so needed in a reactor can also be used further to raise the concentration to the higher concentrations needed to fashion a crude nuclear explosive. Plutonium, chemically extracted from spent fuel rods, can either be plugged back into the reactor as fuel, saved for later use in breeder reactors, or used in a nuclear weapon. Civil nuclear power provides the strategic benefit of freeing oil, which might otherwise be used in electricity generation, for use in military equipment. So much has been written about the strategic benefits and drawbacks of nuclear weapons that it would be senseless to rehearse the litany here.[2] In brief, the use or the threat of the use of nuclear weapons can be used (1) to deter either conventional or nuclear attack, (2) to bolster military claims against others, or (3) to assert regional hegemony or great power status. As related to conventional forces, nuclear weapons may be considered either a supplement, to be used only in the last resort when all else has failed, or a substitute, to compensate for conventional weakness or even to reduce defense budgets. (Nuclear weapons, once developed, provide more 'bang for the buck' than do conventional forces.) The possession of deployable nuclear weapons may reflect either defensive or offensive purposes. It can provide the ultimate deterrent for a nation whose very survival is at stake, such as Israel, South Korea, or Taiwan. It could attract irresponsible leaders, such as Idi Amin or Muamar Qadhafi, eager to gain international attention or to pursue visions of self-aggrandizement and territorial gain. Nuclear weapons may be desired for other aggressive but less capricious designs, or to alter alliance relationships. If Brazil had them, for instance, a Brazilian–Chile alliance could seriously alter Argentine calculations in the traditional border disputes with Chile.

The very suggestion that a government is seeking to acquire nuclear weapons, whether grounded in fact or fiction, can also be used to increase state security. Governments can exploit such suspicions to rally domestic support or to bargain toughly with nations opposed to the proliferation of nuclear weapons. So long as these governments deny any military intent, they can at least deflect international opprobrium. That is why the Indian government called its nuclear test 'peaceful', though the important implications were military. Argentina's sedulous development of nuclear technology and resistance to international safeguards fostered the appearance that it might be developing nuclear weapons. This appearance, however (intentionally) misleading or exaggerated, enhanced Argentine political influence. The governments of Israel, Pakistan, South Korea, and Taiwan have also been suspected (with varying degrees of certainty) of seeking nuclear weapons, though none ever announced that this was its policy. The amount of leverage will differ according to the probability of the allegations and to whether or not a program is consummated by a test explosion.

There are many steps between crude fission explosive and more wieldy and powerful weapons. The ability to deliver the weapon, by truck, bomber, artillery, or ballistic missile, also affects the political impact of their existence. No precise assessment of a value for any of the many rungs on the ladder of nuclear capability can be derived with confidence. Any assessment of the danger that a nuclear weapon might in fact be used must account not only for ability, but also for likely intentions and threats confronted. A government threatened with extinction or led by a megalomaniac would be more likely to resort to the use of weapons, which could ultimately prove suicidal to the user.

Whether a government chooses to commence a military nuclear program depends partly upon what response its officials expect from friends and foes. A Third World government might hope to impress others and to prompt the nuclear weapon states to limit or reduce their own nuclear arsenals in order to discourage proliferation in more countries. As one African scholar said, 'today no nation is respected which does not possess nuclear weapons'.[3] A threatened government, even one lacking obvious nuclear targets, such as South Africa, may wish to convey to its opponents an image of potentially cataclysmic desperation. Such a government evinces a prospect of insanely apocalyptic solutions to its security problems, which cannot lightly be dismissed. Military nuclear programs also have domestic audiences. Public opinion in India has often strongly supported nuclear weapons efforts, as have numerous analysts and officials there.[4] The Pakistani response to the Indian nuclear program – 'what's good for the goose is good for the gander', in Bhutto's words – plainly was aimed not only at alarming the Indians but also at currying public favor at home.

Nuclear weapons and nuclear power stations both entail drawbacks, too. For a country with a limited scientific, industrial, and financial base, the resources devoted to a nuclear weapons program could be used for other weapon systems which are more likely to be used and hence to provide a credible deterrent. The greater the certainty of nuclear weapons possession, the greater are not only the military benefits, but also the political costs. Allies may be alienated and neighbors provoked, regardless of whether or not the proliferator has adequate means to deliver his fission weapons. The Indians paid a heavy price for their Pokharan test: the termination of the Canadian nuclear assistance which had ultimately made the test possible. (The CIRUS research reactor from which the fissile material was obtained was the product of North American assistance.) The US response was to hold back shipments of enriched uranium fuel from the Tarapur station, thereby forcing the TAPS to operate well below peak capability.

In 1981, the Reagan Administration notified Delhi that it could no longer honor the 1963 agreement to supply low enriched uranium to Tarapur, in the absence of Indian ratification of the NPT or acceptance of full-scope safeguards. When the Pakistani nuclear effort took an obvious turn to the quest for unsafeguarded nuclear fuel cycle facilities, the North Americans terminated all nuclear assistance, leaving KANUPP hobbled, lacking adequate parts and fuel. Under the 1976 Symington Amendment (since repealed), the US government also cut off $400 million in military and economic aid promised to Pakistan, leaving only $15 million in food assistance. (Military aid was resumed by President Reagan, in response to the Soviet presence in Afghanistan.) The important linkage between security guarantees and nuclear proliferation can also be seen in the Korean case. The strongest indications that the South Koreans might take up the nuclear weapons option coincided with President Carter's announcement of the pending withdrawal of American troops from the peninsula. To the extent that great power security guarantees are maintained with confidence they offer one of the strongest deterrents to nuclear weapons development, which could compromise a government's vital allied support without substituting an equally powerful indigenous force, able to confront conventional, let alone nuclear, threats.

Indeed, a country seen to be seeking nuclear weapons could doubly jeopardize its security, provoking potential enemies to develop nuclear weapons or to attack preemptively with conventional forces, while at the same time sacrificing allied protection against these foes. Indian Prime Minister Desai suggested that he might order a surgical strike against the Pakistani nuclear effort. His successor, Mrs Gandhi, told Parliament that the government would take appropriate action in response to Pakistani nuclear efforts. The core of the Osirak reactor was blown up by saboteurs at the French port of La Seyne-sur-Mer, near Toulon, in April 1979. In the first few weeks of the Iran–Iraq War, the Iraqi nuclear center housing Osirak was bombed from the air, and in June 1981 was destroyed in an Israeli air raid, confirming the traditional notion that nuclear weapons programs may become military targets before they can be militarily useful. The Israeli precedent must affect the calculation of whether such an escalation of the strategic stakes is dangerous or justified. Lastly, nuclear reactors themselves may become targets – as they already have for environmentalists – for terrorists or foreign adversaries, who could create either chaos and terror, by plunging strategic regions into darkness, or carnage, if they possess the skills to release radiation to the environment. (Any centralized electrical system would be vulnerable to such power cuts, regardless of the energy technology used.)

The overall effect of the actual introduction of nuclear weapons into a region is difficult to predict. If Argentina acquires nuclear weapons, it might become the dominant power of Latin America. On the other hand, Brazil would probably follow suit. The two nations might find that their security had increased, if they reached a strategic balance in which each deterred the other from attack. Alternatively, the two new nuclear weapon states might find that their security had diminished, as they faced increased stakes (wholesale destruction) with increased uncertainties (both in ensuring accurate warhead delivery and in judging the opponent's capability and intentions).

Similar uncertainties characterize other regions. Should Pakistan test a device, India could well resume its 'peaceful nuclear explosive' testing program, or even advance its military nuclear capabilities by improving its bomber or missile capabilities to deliver Indian-manufactured nuclear weapons. A whole range of strategic postures could be assumed in the subcontinent, categorized as follows by Richard K. Betts:

- Deterrence by uncertainty. Maintenance of any level of nuclear force, perhaps with policy on use unarticulated, with reliance on the enemy's unwillingness to test the defender's will to use the weapons recklessly and the enemy's lack of confidence in its own first-strike capabilities.
- Proportional deterrence (à la Gallois). Maintenance of a force capable of targeting population, assets, or forces, the destruction of which would exceed in cost the prospective gains from aggression.
- Superiority. Ability to inflict greater damage on the adversary than it could inflict on the defender, preferably even after absorbing a preemptive attack.
- 'Assured heavy damage.' Similar to the American conception of assured destruction, but without the certainty that the damage inflicted would thoroughly cripple the target country's society or economy. The damage would clearly be formidable but not 'unacceptable' beyond doubt.[5]

The effects of the introduction of nuclear weapons into the Middle East are even harder to predict, and could be even more disastrous, due to the volatile admixture there of Israeli isolation, Arab radicalism, Islamic fundamentalism, and (sometimes state-sponsored) terrorism. The word 'introduction' is used here because, despite widespread belief that the Israelis already possess nuclear weapons, they themselves have always insisted that they would not be the first to 'introduce' nuclear weapons to the region. An overt South African test would intensify its embattled status, and could well push black African nations, notably Nigeria, into accelerated nuclear development.[6]

POLITICAL USES

Second, because of the especially sensitive nature of nuclear technology, even a country with patently nonmilitary intentions can exert political leverage with a civil nuclear program. International regulation of nuclear technology is extraordinarily strict, because even rudimentary activities can improve weapon-building capacities. Desires to monitor peaceful nuclear activities worldwide has led most nations to cede enough sovereignty to the International Atomic Energy Agency to permit the application of safeguards, in the form of facility design requirements, material accounting requirements, inspections, and other methods of material control. As demands have grown for stricter safeguards, some governments have resisted, on grounds that additional controls go too far in breaching sovereignty. Of course, this resistance can be used for bargaining.

A government can repeatedly assert that its aims are benign, but reject nonproliferation accords as discriminatory against countries barren of nuclear weapons, since nuclear arsenals are permitted in a handful of states. The NPT, complained CNEA President Castro Madero, 'disarms the unarmed, and allows those with nuclear weapons like the United States and the Soviet Union to keep them'.[7] Brazilian policy exemplified this attitude: the government accepted international safeguards over all nuclear facilities in the country, ratified the treaty which banned nuclear weapons from Latin America, and attested its disinterest in nuclear weapons, but refused to sign the NPT.

The failure of the nuclear powers to abide by their obligations under NPT Article VI, to negotiate in good faith on effective arms control and disarmament measures, serves as a lightning rod to developing country discontent. The nuclear weapon states are chastised for permitting the untrameled expansion of their own nuclear arsenals (vertical proliferation) while professing undue concern for the spread of nuclear weapons to more countries (horizontal proliferation). Even though the notion that vertical and horizontal proliferation are intimately linked is a shibboleth – the decision to go or not to go nuclear depends far more upon a government's calculation on how its own position would be strengthened or weakened than upon the immorality of the great powers – it provides a powerful club with which to batter the Americans and Soviets. In 1980, controversy over Article VI and the nuclear weapon states' refusal to accede to a specific, multiplank arms control directive proposed by delegates from the Group of 77 developing countries caused the second NPT Review Conference to adjourn in discord, without any final declaration to show for its month-long negotiations and much longer preparations.

Argentine governments have masterfully tread the line of studied ambiguity. On its face, the record is clear and consistent. Every leader of the Argentine nuclear program affirmed its exclusively peaceful intentions. Although the hapless Ronald Richter boasted in 1951 of Argentina's ability to build hydrogen weapons, he also claimed that President Peron repeatedly rejected the idea of a nuclear weapons program. Quihillalt declared straightforwardly, 'We never thought of making a nuclear device in the country.'[8] In his second term as CNEA chief, Admiral Iraolagoitia stated that, 'the country has no project for, nor conditions to develop, an atomic bomb'. His deputy, Mario Bancora, added, 'I suppose we can expect there will be some advocates for nuclear devices in Brazil and Argentina, but we can only hope that nobody gives any credit to such nonsense here or there.'[9]

On the other hand, Argentine leaders refused to accede to any international commitments that were deemed discriminatory, such as the NPT. In their view, the Treaty not only codified an invidious distinction between nuclear and nonnuclear weapon states, but also compounded the injustice by compelling only the latter to submit all nuclear facilities to international safeguards. During the 1968 debate on the NPT, the Argentine representative stated that his government could not 'accept remaining subordinate to a continuing dependence on the great powers for nuclear technology, especially when our country has laid the foundations for a nuclear technology needed for economic development'.[10] In the years following the Treaty's 1970 entry into force, nonnuclear weapon states found more reason to refuse to join, condemning the nuclear weapon states party to the Treaty (the United States, Great Britain, and the Soviet Union), for their failure to fulfill their obligations under Articles IV, V and VI. These articles assured the nonnuclear weapon state parties that the nuclear weapon states would facilitate 'the fullest possible exchange of equipment, materials, and scientific and technological information for the peaceful uses of nuclear energy', would make available the 'potential benefits from any peaceful applications of nuclear explosions', and would 'pursue negotiations in good faith on effective measures relating to the cessation of the nuclear arms race at an early date and to nuclear disarmament'.[11]

Argentina also refused to join every other major Latin American country (except Cuba) in ratifying the Treaty of Tlatelolco, which proposed to ban all nuclear weapons from Latin America, because it could not accept the US interpretation of the Treaty's definition of 'nuclear weapons' as also prohibiting 'peaceful nuclear explosives', since the two were indistinguishable from one another.[12] A decade after the Treaty's submission, the Argentine delegation to the UN Special Session on Disarmament finally promised to ratify Tlatelolco. Four years later, this pledge remained unredeemed.

Doubts concerning Argentine sincerity in denying any nuclear weapon objectives also arose among those who regarded the discrimination issues as being a pretext, used merely to avoid acceptance of binding commitments to forswear nuclear weapons. Occasional CNEA and press statements that 'we could if we wanted to' reflected a proud and militant attitude. Critics recalled that Argentina had signed a five-year agreement with India to cooperate in peaceful nuclear research, only ten days after the latter had tested a nuclear explosive beneath the Rajasthan Desert.[13] Though Argentine claims that the Indians had requested the end of May signing date months before were plausible, the lack of any apparent effort after the Indian test to reschedule it indicated that perhaps the coincidence did not displease the Argentinians.

Further grounds to doubt Argentine sincerity in renouncing nuclear weapons, the NPT, and Tlatelolco alike, was provided by the government's avoidance of safeguards wherever possible. Progress on a renewed safeguards agreement for the Atucha-1 reactor was so slow that the CNEA faced the possibility of reactor shutdown pending agreement.[14] The Atucha-2 negotiations seemed headed toward a contract, based on German-Canadian agreement, that would require Argentina to accept IAEA safeguards over all Argentine nuclear facilities present and future, as required of NPT nonnuclear weapons state parties. When the CNEA divided the package into separate competitions for the heavy water plant and the Atucha-2 power station, the Germans stole a march on the Canadians, selling only the reactor and dropping the requirement to accept NPT-type safeguards. Afterwards, the CNEA still balked at safeguards insisted upon by the German government, to the point where speculation arose that the deal might collapse. While this resistance might have been based solely on the allegedly discriminatory nature of safeguards, it generated sincere doubts about Argentine nuclear intentions.

Prestige is another important political factor. Enhanced respect is always welcome. Large projects often appeal to developing country governments as a means to demonstrate their ability. Because of its complexity, perhaps even its mystery, the mastery of nuclear technology can instill popular pride as well as enhance the legitimacy of a central government eager to contain restive partisans. Admittedly, prestige can serve many purposes unrelated to state security, such as gratifying leaders' egos. In the light of the political strife endemic in the developing world, though, it is safe to assume that, to the extent that prestige is important in nuclear policy-making, it is desired at least partly to improve state security. This assumption is often borne out by the way in which nuclear policies are presented. Some leaders frankly concede that their primary concern is to stay abreast of foreign developments or to gain international influence, and have little to add when discussions turn to the specific uses planned for the electricity to be generated in their nuclear power programs. Others are extremely vague or extravagant in speech, suggesting that their words, apart from the actions proposed, are intended to elicit a favorable response. Where the planned uses of nuclear-generated electricity are obscure, the reason may be that electricity production is not the essential concern. This does not prove that the nuclear stations are desired simply for prestige purposes, for other goals (such as energy diversification or technology transfer) may also be important. Still, in such cases, prestige motives may well be involved.

The prestige benefit from nuclear programs can be claimed by almost any government. It does not sharply discriminate between the able and the inept, at least partly because prestige can be self-defined. A general who boasts that he is undefeatable is presumed to be stronger than one who complains that his forces are weak, before one examines the forces each leads. Similarly, Peron enjoyed some prestige, albeit briefly, from his absurd claim to have harnessed the fusion process. His announcement of Richter's experiment could have been used to distract attention from his other domestic political problems. The desire to recoup the prestige lost through Richter helps explain Peron's anomalous desire to hire competent scientists, even if they opposed his policies. The determination to demonstrate independence has always displayed a strong public relations aspect, as extensive press coverage indicates.

The traditional competition for influence in Latin America undoubtedly contributed to the push behind nuclear energy in both Argentina and Brazil. Nuclear technology is one of the few fields in which the Argentinians traditionally lead the Brazilians, and this advantage extends to the first efforts of each nation to enter the nuclear export field. Brazilian President Geisel's decision to conclude the largest nuclear export agreement in the world at least partly reflected an effort to redress its inferiority. Still, mutual interests could at times overshadow differences, as in May 1980 when the Argentine and Brazilian foreign ministers pledged mutual assistance in atomic development.[15]

Sukarno, too, asserted his intention to test nuclear weapons without any likelihood of fulfilling that pledge. In February 1965, Brigadier General Hartono, director of the Army Supply Service, said that Indonesia would be able to produce atomic weapons, as well as guided missiles, that year. Western experts countered that the Indonesians lacked not only the scientific expertise, but also the industrial capability, financial resources, and fissile material to manufacture a device. But the boldness of Sukarno's rhetoric, even suggesting possible test dates, provoked second thoughts. Perhaps he would detonate a device consisting of radioactive pellets incorporated into a conventional burst bomb. Some speculated that the Chinese would assist Sukarno or even agree to test one of their devices in Indonesian territory.

These alternative hypotheses were equally implausible. Few believed that the Chinese, stretched to carry out their own nuclear weapons development program without Soviet assistance, would be able to help others. Many doubted that, despite their radical public statements, the Chinese would feel more secure with the finger of President Sukarno, who had recently boasted of missiles 'which could destroy Singapore and Kuala Lumpur', on the nuclear trigger.[16] Besides, the Chinese might not have wished to conduct a test in Indonesian territories, where it could be more easily monitored by the Americans. Sukarno's regime did not survive the fateful year, leaving historians to speculate what, in fact, he intended to do. There was, of course, no nuclear weapon test.

Sukarno's commitment to nuclear power supported his commitment to national unity. The ability to harness the atom for scientific and eventually commercial purposes was seen to measure advancement and independence, relative to the industrialized countries as well as to Indonesia's developing neighbors. Pride in this ability perhaps would encourage Javans, Sumatrans, and Sulawesans alike to tender more allegiance to the Republic, and less to their own islands. If uranium deposits were found on Sumatra, Kalimantan, or Irian Jaya for use in stations in Java, the increased interdependence among the islanders could strengthen the national bond.

Suharto's nationalism also recognized the prestige value of nuclear development, as reflected in his comment in response to the May 1966 Chinese nuclear test that Indonesia did not wish to fall behind in the development of nuclear technology.[17] But the prestige from an active nuclear power or weapon program has not been essential to Suharto's nationalism. The image of an energetic nuclear effort as a trophy of independence has been abandoned as too expensive and complex to pursue. Suharto could relinquish this symbol without political cost, not only because of the wealth of alternative energy resources in Indonesia, but also because of the tendency of the Indonesian political elites to follow any presidential policy or its reversal, so long as resources are supplied for their personal wants. More than in most political systems, the personal beliefs of the chief executive in Indonesia leave their stamp on policy outcomes large and small.[18] By the late 1970s, Suharto's personal coolness toward nuclear power development was reinforced by the growing unpopularity (hence loss of prestige) of nuclear power worldwide. Philippine President Marcos in 1979 suspended work on the only power reactor under construction in Southeast Asia, criticizing the Westinghouse unit as poorly designed and potentially unsafe. This action deflated most of the prestige value which still attended the idea of a nuclear power station in Indonesia.

The Shah openly linked his plan with increased Iranian prestige and influence. When announcing the introduction of nuclear power to Iran, in his Now Ruz (New Year's) speech of March 1974, the Shah declared: 'We have not lost our heads. But, all the same, the nation should know that Iran enjoys a prestigious world position which is counted on increasingly.'[19] He promoted nuclear power as part of an imperial plan to make Iran 'one of the five great non-atomic weapon countries of the world within a single generation'.[20] The mechanism of this transformation, however, remained unclear, since not even AEOI officials seemed to know exactly to what use the installed nuclear power at Bushehr and Ahwaz would be put, though it was stated that desalination would not be the only purpose.

Leverage, unlike prestige, cannot be obtained by fiat or bold assertion, because it depends upon the acquiescence of a target. If that target is insensitive to the nuclear policies of another government, then it cannot be subject to leverage. Consequently, leverage in the nuclear field rises in parallel with capability. It follows that countries such as India and Argentina should enjoy the greatest nuclear-related leverage among developing nations, and those such as Malaysia and Nigeria the least. This is, in fact, the case. Leverage takes two forms. The first is direct threat, or trade, depending from which side one views the transaction. For instance, the Indians threatened to reprocess plutonium from American-supplied enriched uranium fuel if the United States abrogated its 1963 Tarapur agreement. This argument carried great weight, and prevailed for seven years after the 1974 Indian nuclear test, precisely because India had already demonstrated its ability and willingness to obtain and detonate plutonium.

The second form of leverage is the discrimination allegation. Politically, countries such as Argentina, Brazil and India have been the most skilled exploiters of the advanced nations' sensitivity to the acquisition of nuclear technology by Third World nations. The weight of their allegations of discrimination is a function of their technological capability. The CNEA, Nuclebras, and the DAE are advanced enough to argue that they could benefit from technology which some industrialized countries refuse to supply, and that this denial has seriously damaged efforts to obtain self-sufficiency in the nuclear field. Yet they are not so far advanced as to preclude the claim that, as Third World countries, they are entitled to technology transfers as a matter of right. By contrast, allegations of discrimination from Zaire or the Netherlands would lack cogency, due to the former's inability to use technology effectively were it to be supplied, and to the latter's status as a nation so advanced that its 'entitlement' to others' technology would not extend beyond fulfillment of the terms in a commercial arrangement. Within the bounds of its overall resources and technological capability, each government influences its own political clout, by the amount of effort it devotes to nuclear development. Governments deeply concerned with demonstrating their independence, such as India, Argentina, and South Africa, tend to be those which are (1) most committed to independent nuclear development, and (2) most reluctant to alter their course except in exchange for major concessions. By contrast, those countries closely allied to the industrialized nations tend to fall into the dependent nuclear category and its more malleable ways. This deprives them of a great deal of leverage. A larger nuclear program increases political leverage for both buyer and seller. The country purchasing the reactors can use the carrot of numerous sales contracts to wring concessions from suppliers. In the mid-1970s, this policy enabled Brazil, Pakistan, and South Korea to induce France and West Germany to sign contracts for plutonium reprocessing facilities.

Leverage works in the other direction, too. This fact tended to be obscured in the late 1970s, when the depressed nuclear market seemed to give buyers the stronger hand. Though it is sometimes hidden, the leverage of the industrialized nuclear supplier nations over their developing country customers cannot be erased. Even India and Argentina, the two strongest nuclear independents, have both been forced to yield to supplier pressures. The Argentine case will be elaborated in Chapter 10. In India, one analyst chided the Indian government for its 'growing submissiveness', citing the decisions to open Tarapur to IAEA inspection, to extend Canadian inspection for RAPS-1 from the first generation fissile material to IAEA inspection during all subsequent generations, and to continue to permit Canadian and IAEA inspection of RAPS even after Canada's 1974 termination of all nuclear assistance to India.[21]

As noted, developing country leverage increases with the sophistication of its nuclear program. This leverage is far less reciprocal than that provided by a large effort, because some developing countries can develop dangerous capabilities independently, but none can build a large nuclear power program independently. In fact, research-scale programs often pose the most serious and least controllable nuclear weapon risks, as witnessed by the worldwide concern surrounding Israeli, Iraqi, Indian, and Pakistani research efforts. In order to maximize their political leverage, without compromising their professedly peaceful intentions, governments in all of these countries (though Israel to a lesser degree) refrain from avowing any nuclear weapon intentions. Such an admission would cost them a good deal of the leverage which they so carefully have marshaled.

Politically, nuclear power also can provide domestic security benefits. The demonstration of technological advancement involved in nuclear research can provide the prestige eagerly sought by leaders who wish to shore up domestic support. More tangibly, nuclear power can help consolidate and extend government authority throughout the economy, insofar as it is imposed upon businesses which theretofore had relied on their own furnaces and electrical generators. Rural electrification schemes serve the same purpose, forcing the populace to depend on the leadership for an essential service, while widening the base for government revenues to include all buyers of state-generated electricity. Governments may manipulate rate structures to reward friends, penalize foes, and tame possibly unruly elements (by providing underpriced electricity to rural areas, for example). This subjection cannot be wantonly exploited, since it involves serious risks. The all too frequent power cuts in Iran in the mid-1970s suggested government ineptitude, contributing to the serious domestic unrest which in July 1977 unseated the thirteen-year-old Hoveyda government. Generally, popular discontent represents the greatest political liability of nuclear power programs, and may arise through increased environmental concerns, aggravation at the high costs or other failings of nuclear power, or government inability to fulfill its professed nuclear objectives. One other political drawback is the international disapproval engendered by certain nuclear programs. Punishment is by no means evenly meted out, for pariahs like Israel and South Africa are most often excoriated by the Group of 77, while India and Pakistan are usually subjected to pressures from the West to curb their nuclear activities. Nevertheless, all four of these governments suffer in some way or another for the political gains derived from their relatively advanced nuclear status.

ENERGY SECURITY

Third, nuclear power can be sought to provide energy security. Serious disagreement exists over whether energy should be considered an ordinary economic good, or a factor so fundamental and irreplaceable that neoclassical economic analysis does not apply. Theoretically, the distinction between energy security and other economic issues is artificial, because economic security costs and benefits ought to be quantifiable. Nevertheless the concepts of economic and energy security have such salience in their own right that they merit some separate discussion.

Argentine sentiments are typical, and help explain what appears to be occasional economic irrationality. Resentment cultivated over a century of dependence on (and perceived exploitation by) foreigners in the energy sector has sensitized governments to energy as a security matter, which may require short-term economic sacrifice. Political considerations in the early 1960s moved Argentine governments to tap domestic oil reserves in order to reduce dependence upon imports, even though cheap oil was abundantly available from the Middle East. Though the country is still 90 percent self-sufficient in oil supplies, Argentine economists wince at the expense of the shortfall which must be made up in imports, complaining that 'if we had been less nationalistic then, that oil would still be under our feet today'. In the nuclear field, a *Clarin* editorial argued that financial considerations were not overriding factors in the 1979 decision to proceed with the fifteen-year, four-reactor plan. 'At play', it said, 'are policies of national security and sovereignty, energy problems, the definition of economic and technological development plan [sic].[22] The desire for energy security helps explain the apparent economic illogic of choosing the more expensive KWU bid for the Atucha-2 station. The Canadians, at Embalse, had used threats of work stoppage to manipulate Argentine safeguards policy and to extract more in payments. This rankled. The German firm could be relied upon to stick to the letter of an agreement and not to engage in such arm-twisting. The Atucha-2 competition showed that this factor could be decisive.

The oil crisis taught governments that energy supply transcends other economic issues in importance to state security. Both exporters and importers of oil seek to reduce its domestic consumption – the former to maximize profits, the latter to minimize expenses and vulnerability to cartels. In Iran, for instance, oil reserves are so limited that, by the end of the century, the country may well have ceased to be a major oil exporter. The Shah wished to maximize the benefits his country could reap from these resources by extending them as long as possible, especially through encouraging their use in the petrochemical production rather than the electricity generation sector. He also wished to transform the temporary influence given his country by its petrodollars into a more lasting influence based on the wise investment of these revenues. Nuclear power seemed the best way to obviate the need to consume oil at home, while using export receipts of the day to assure energy security for the next century. The continuing importance of oil in a world where it would become increasingly expensive seemed destined to cause insecurities and perhaps crises, if substitutes were not soon developed. The Shah hoped that his nuclear program would guarantee Iran's energy future, and thereby place it in an advantageous position of increased world influence in the decades ahead.

For oil importing nations, the key security benefit derives from the diversification of energy resource dependence; fission offers an escape for excessive dependence on petroleum imports, and thus could reduce vulnerability to OPEC actions. The cheap, reliable provision of electricity hoped for in nuclear plans could also enhance security in the event of an economic attack from any quarter. To the extent prosperity reinforces stability, energy policy is critical, as is the role of nuclear within it. On the other hand, increased dependence upon (hence vulnerability to) foreign suppliers of nuclear technology, and perhaps natural or enriched uranium, could imperil security, as much as OPEC ever did. It is already clear that supplies of nuclear technology and materials are subject to the vagaries of changing political situations in supplier countries. So whether adoption of nuclear power on balance improves or worsens a developing country's energy security depends upon how its relations with OPEC compare to those with particular nuclear suppliers.

The energy security benefits are most concentrated for those nations heavily dependent upon imported oil for electricity generation. Obviously, nuclear reactors cannot relieve petroleum demand for transport uses. The harm flows from the inelasticity of oil demand and its low substitutability in the short run; consumption and production patterns cannot be transformed overnight. An economy serviced by a truck network cannot quickly reduce diesel fuel consumption. New energy technologies and economic infrastructure must be developed. Take, for instance, the invention of a new light bulb which used less than half of the normal light bulb electricity requirement. If that bulb could only fit a square socket, round sockets would have to be replaced, both delaying and reducing the anticipated energy savings. Where oil embargoes or price increases force major sacrifices, security may be compromised, for a suffering economy grows prone to intrusion, subversion, or coercion.

POLICY DETERMINANTS

The influence of security objectives upon nuclear policies cannot be understood merely by cataloging the pros and cons of various approaches with respect to the military, political, and economic dimensions of security. Choices are also shaped by technological capabilities, the nature of a particular regime, official policies, and outside constraints. Examination of these additional factors helps clarify the role of security objectives in nuclear policy-making.

The extent to which technological capabilities determine nuclear policies is partly, but only partly, a matter of choice. A government may choose to mount a relatively large or small nuclear technology effort. It may accede to the NPT in order to avail itself of the promised 'fullest possible exchange' of nuclear technology for peaceful purposes. Other governments may see submission to the NPT (and so to its great power authors) as abject truckling, unworthy of the technology offered in return, a view reinforced by the nuclear weapon states' party to the Treaty refusal to offer certain technologies, such as reprocessing facilities and peaceful nuclear explosions. Which of these two opposing views prevails in a government depends, among other things, on whether nuclear power or nuclear independence is more desired.

'Sensitive technologies' are those which facilitate the acquisition of fissile material suitable for a nuclear explosive. Initially, only plutonium reprocessing and uranium enrichment facilities were included in this category, but as nonproliferation concerns grew, the definition was broadened. Heavy water reactors are good plutonium producers, so heavy water production facilities were included on the Suppliers' Group list of exports which should trigger the application of safeguards. Once a country became self-sufficient in nuclear technology, leverage against it would be minimized and one no longer could ensure that future facilities would be submitted to IAEA safeguards. Consequently, any step that could complete a nation's fuel cycle capabilities became in that sense sensitive. More directly, technologies that could convert a crude fission device into a deliverable nuclear weapon – metallurgy, chemical engineering, aircraft and missile construction – enhance the security benefits available to a government.

Different fuel cycles can be chosen. Generally, the heavy water fuel cycle can more easily be applied to weapons production than can the light water cycle, because (1) weapons-grade plutonium can more easily be obtained than can weapons-grade uranium, and (2) weapons-grade plutonium can more easily be obtained from a heavy water than from a light water reactor. A heavy water reactor is a better plutonium source because it uses natural uranium, which has higher concentration of the uranium-238 which is transformed into plutonium, and because fuel rods can be extracted while the reactor is in operation, permitting both discreet and early removal. Bomb makers prefer early fuel removal, because extended irradiation bombards more plutonium-239 with neutrons and therefore creates, through neutron absorption, more plutonium-240, a non-fissile isotope which 'poisons' plutonium as an explosive.

Generally, then, countries that choose heavy water technology and develop fuel cycle self-sufficiency are those most able to exploit the military and political leverage advantages of atomic energy. The Indian reprocessing plant at Trombay yielded the plutonium used in the Pokharan test. Once the French withdrew from the Chasma project, Pakistani reprocessing efforts continued, but emphasis shifted to the Kahota enrichment project. South African development of independent enrichment capability facilitates the weapon option. Israel's Dimona center contains a natural uranium research reactor and is believed to provide full fuel cycle capabilities. The weapons industries of India, Israel, and South Africa elevate these three countries to the most militarily threatening of the developing country nuclear ranks.[23]

The heavy water path brought nuclear weapons within technical reach in Argentina. Castro Madero has asserted that Argentina is able to build nuclear weapons.[24] The man who headed the CNEA technology division for many years under Quihillalt said that Argentina could build a bomb in four years 'at a very reasonable cost – say $250 million, which is ten months' deficit on the Argentine State Railways.'[25] One outside analyst confirmed that the Commission was probably capable of producing a crude nuclear device, using plutonium produced in its research reactors, as secretly and as easily as heroin could be produced.[26] Central to this capability was the government's choice of the heavy water fuel cycle. A special feature of the CANDU reactor (being built at Embalse) was that the fuel rods, interspersed in a matrix of pressure tubes rather than encased by a pressure vessel, could be withdrawn without shutting down the reactor. Fuel rods containing fissile plutonium could be removed surreptitiously from the Embalse CANDU reactor at any time after it began operation, and the diversion could be scheduled for the least conspicuous moment. Fuel rods could only be removed from Atucha-1 at the obvious time of a reactor shutdown.

The greatest danger for the military use of the program was that weapons-grade plutonium could be extracted from irradiated fuel rods through reprocessing. The CNEA built and operated a small reprocessing plant in the 1960s, which probably produced enough plutonium for some nuclear explosives.[27] In 1979, Castro Madero announced that the Commission would begin to 'produce' plutonium on a much larger scale, in a new reprocessing facility.[28] Perhaps more than anything else, the prospect of the CNEA extracting significant quantities of plutonium in a home-built, unsafeguarded facility evoked concern within the nonproliferation community.

Some of the nuclear dependent countries – South Korea, Taiwan, and Brazil – also possess the technological maturity to enjoy political clout from nuclear technology, but cannot press the point too far without displeasing the suppliers, upon whom they depend and toward whom they are legally bound, through safeguards and other legal connections. Iranian technological capabilities did not exclude the possibility of nuclear weapons. Research on the development of laser isotope separation, a nascent uranium enrichment technology that would greatly facilitate the manufacture of weapons-grade uranium, was conducted in AEOI laboratories. A radioisotope laboratory and its technicians were hived off for use by the military, mainly for development of nighttime signaling techniques and other nonweapon uses, according to AEOI officials. Yet capabilities developed in the civil nuclear program could not have had nuclear weapons as their primary objective. First, technology transfer was low priority, in contrast to the Argentine program. AEOI officials felt that they did not have time to waste on research reactors before going commercial. Though nuclear technology transfer was the subject of a major international conference hosted by the AEOI, homage was paid these transfers more in principle than in practice. Top technical priority at the Organization was given, not to research, but to training the thousands of reactor operators and maintenance personnel who would be needed to keep the reactors in service. A government interested in nuclear weapon development would have laid far greater emphasis upon developing scientific acumen, so that its own experts could learn techniques that others would not teach them. A nuclear research program would have posed a graver proliferation risk than *any* electric power program, let alone one with such minimal technology transfer.

Second, if the objective of the Shah's nuclear power program was military, then he picked the wrong technology. The advantages of the natural uranium fuel cycle for nuclear weapon production have been described. Heavy water reactor orders could have been justified easily at the 1974 inception of the AEOI, since at that moment a shortage of uranium enrichment capacity for light water reactors was widely predicted. Nevertheless, the Iranian government opted for light water reactors. No effort was made to acquire domestic enrichment facilities. Resistance to restrictions on reprocessing rights for spent fuel was based on the discriminatory nature of such constraints, a complaint that entertained support in Europe and Japan. The Shah told visiting American Congressmen that he would gladly forgo reprocessing in return for guaranteed reactor fuel supplies. Iran eventually yielded on the reprocessing issue. Of the nonnuclear countries, Iraqi capabilities and possession of weapons-grade uranium cause the greatest nonproliferation concern.

Official policies are a second major litmus to help measure the nuclear energy linkage to security. Capability can reveal only half the picture; technology alone is neither benign or malign. Only when wedded to political will can it provide leverage or prestige of any sort. To the extent that prestige is self-proclaimed, the contribution made to it by nuclear programs is also flexible. The Indian government has always declared the importance of nuclear power to its independence; the Moroccan government has not. This distinction does not imply that one government is more independent than another, or that a nuclear program contributes to national independence more in one place than in another. It merely reflects a difference in policy. This policy can quickly change events in one country, as when Suharto dropped his predecessor's emphasis on nuclear achievement as a building block of nationalism. Harsh rhetoric and resistance to compromise are least available to the nuclear dependent countries, which have the most to lose from alienating suppliers.

Governmental statements provide concrete evidence of the relationship between nuclear policy and security objectives. Formal pronouncements invariably aver exclusively peaceful intentions. India has always maintained that its Pokharan test was peaceful in aim, with possible applications to excavating harbors, cutting roads and railways through mountains, building canals and reservoirs, and extracting inaccessible fuels and minerals.[29] Prime Minister Gandhi wrote a 57-year-old survivor of the Hiroshima nuclear attack that: 'On the day the bomb was dropped over Hiroshima, my father, Jawaharlal Nehru, called it the "death-dealer" and since then, we have stood for complete disarmament under effective international control . . . . India does not possess nuclear weapons and has no intention of developing or producing them.'[30]

Governments suspected of developing nuclear weapons often are especially circumspect in their statements about them, due to apprehension that others might react unpleasantly. Argentina and Pakistan protest their nuclear innocence, but their technical capabilities and political situation undermine their credibility. Following their deployment of nuclear weapons, Chinese statements that such devices generally benefited nations that acquired them gradually ceased. The world community is hypersensitive to nuances of nuclear policies in technically capable nations, but insensitive to those in the least technologically developed countries. Consequently, countries like Indonesia need circumspection less, and indeed can only gain worldwide attention through outlandish statements of an imminent nuclear test.[31]

More aggressive statements are usually left to journalistic, academic, and partisan domains. The Peronist newspaper, *Mayoria*, responded to the 1974 nuclear accord with India by publishing a front page picture of a mushroom cloud, encaptioned 'Argentina and the Bomb'. An accompanying editorial concluded that 'the country can achieve the atom bomb within a reasonable period of time'.[32] In 1975, a Peronist deputy introduced Congressional legislation calling on the government to manufacture a nuclear weapon for national defense. CNEA repudiation of the bill, which never reached the floor, did not allay all concerns raised by it. Meanwhile, Genral Juan Guglialmelli, editor of the influential strategic journal, *Estrategia*, noted the potential importance of nuclear weapons for Argentina.[33]

In India, the director of the Institute of Defence Studies, K. Subrahmanyam, forcefully argued for nuclear weapons acquisition, stating that:

> The only way India can keep its options open is to exercise the nuclear option. When we do that, the United States may come to realise that they could not ignore a nation of 700 million with nuclear weapons.[34]

His view followed a long tradition of nuclear weapons advocacy in India, which for many years had been spearheaded by Homi Babha. Subrahmanyam even quoted Mahatma Gandhi as saying, 'I would rather have India resort to arms in order to defend her honour than that she should in a cowardly manner become or remain a helpless witness to her own dishonour.' Of course, governments influence these fora, and sometimes send purposefully militant signals. The deliberate Israeli ambiguity in nuclear policy declarations has been noted, as have Bhutto's exclamations. In spring 1980, Prime Minister Ghandi reiterated her country's right to conduct peaceful nuclear 'explosions or implosions'.[35] Libyan leader Qadhafi reportedly sent an envoy to Peking to ask Chou En-Lai whether he could buy an atomic bomb (Chou refused) and then sent uranium and money to Pakistan to secure a role in that project.[36] After the June 1981 Israeli attack, Iraqi President Saddam Hussein told his Cabinet that:

> any country in the world which seeks peace and security, respects peoples and does not wish those peoples to fall under the hegemony or the oppression of external foreign forces should assist the Arabs in one way or another to obtain the nuclear bomb in order to confront Israel's existent bombs. This will achieve peace regardless of the Arabs' aims and capabilities.[37]

Words alone cannot be relied upon to reveal intentions, for they are sometimes ill-considered, directed to a passing situation (as Suharto's pronuclear weapon remarks in response to the third Chinese nuclear test), and intended to be quickly forgotten. Positions on international pacts and institutions are less ephemeral. Governments which accept great power security guarantees have an interest in shunning nuclear weapons, so long as they desire to preserve that protection. More explicit commitments are entailed in membership in the IAEA, whose statutory objectives include ensuring 'so far as it is able, that assistance provided by it or at its request or under its supervision or control is not used in such a way as to further any military purpose' (Article II). The IAEA safeguards system reinforces that peaceful commitment, by furnishing a means by which diversion of fissile materials to military uses may be detected. In the past five years the North Americans have attempted to strengthen the safeguards system by insisting that all nonnuclear weapon states (NNWSs) accept safeguards on all their nuclear facilities, as is required of all NPT nonnuclear weapon state parties. These 'full-scope' safeguards are intended, at best, to result in the universal application of safeguards and, at least, to punish politically those countries which resist. (All NNWSs with unsafeguarded facilities are listed in Table 7.1.) Until the present nonproliferation regime can be cleansed of its discriminatory taint, however, harshened international regulations will carry little moral force, depriving them both of their persuasive and punitive values.

The IAEA system, born in 1957, was bolstered by the 1963 Limited Test Ban Treaty (LTBT), whose 105 parties have agreed not to conduct nuclear tests in the atmosphere, on land, or underwater. Among the developing countries, Pakistan is the most notable holdout. A comprehensive test ban could prove an extremely effective nonproliferation measure, both by closing the one remaining test venue untouched by the LTBT – underground – and by accommodating Third World pressures to curb vertical proliferation as a step toward reducing horizontal proliferation. This major step toward fulfilling NPT Article VI would vastly improve bargaining positions with India, which historically has linked acceptance of nonproliferation controls to great power ratification of a comprehensive test ban treaty. (Of course, the 'comprehensiveness' of any treaty could be questioned. But, even if the Indian government condemned a pact for, say, permitting tests up to 5 kilotons or expiring in five years, it still would lose the unassailable moral perch from which it castigates the nuclear weapon states today.)

The next major nonproliferation agreement opened to signature after the 1963 LTBT was the Treaty for the Prohibition of Nuclear Weapons in Latin America, concluded at Tlatelolco, Mexico, on February 14, 1967. This is the only nuclear weapons-free zone (NWFZ) completed for a populated area to date, despite recurrent proposals for others in the Middle East, Africa, and Asia.[38] (Nuclear weapons have been barred from Antarctica and outer space.) With the exception of Cuba and Guyana, all Latin American governments have signed the Treaty of Tlatelolco, and all signatories except Argentina

Table 7.1 *Operating Nuclear Facilities not Subject to IAEA or Bilateral Safeguards, as of December 31, 1978[a]*

Country	Facility	Indigenous or imported	First year operation
Egypt	Inshas research reactor	Imported (USSR)	1961
India	Apsara research reactor	Indigenous	1956
	Cirus research reactor	Imported (Canada/ USA)	1960
	Purnima research reactor	Indigenous	1972
	Fuel fabrication plant at Trombay	Indigenous	1960
	Fuel fabrication plant, CANDU type of fuel elements, at the Nuclear Fuel Cycle complex, Hyderabad	Indigenous	1974
	Reprocessing plant at Trombay	Indigenous	1964
	Reprocessing plant at Tarapur	Indigenous	1977
Israel	Dimona research reactor	Imported (France)	1963
	Reprocessing plant	Indigenous (in collaboration with France)[b]	
South Africa	Pilot enrichment plant	Indigenous (in collaboration with FR Germany)[c]	1975
Spain	Vandellos power reactor	Operation in co-operation with France	1972

[a]Significant nuclear activities outside the five nuclear weapon states recognized by the NPT list is based on the best information available to SIPRI. In addition there are laboratory-scale activities such as the reprocessing facilities in Pakistan (first year of operation: 1970) and Egypt and some small-scale fuel fabrication capability in the same country, established with the assistance of Belgo-Nucleare, Belgium. Furthermore, no safeguards agreement has yet been negotiated for the fuel fabrication or reprocessing plants under construction at Ezeiza, Argentina. The fabrication plants will manufacture natural uranium fuel elements using Argentinian uranium.

[b]Assistance by Saint Gobain Techniques Nouvelles.

[c]Co-operation between STEAG (FR Germany) and UCOR (South Africa).

Source: Stockholm International Peace Research Institute, *World Armaments and Disarmament: SIPRI Yearbook, 1979* (London: Taylor and Francis, 1979), p. 314.

have ratified it. The Treaty has entered into force for twenty-two governments.[39] All countries with Latin American possessions (France, the Netherlands, United Kingdom, and United States) have

signed and all nuclear weapons powers have signed and ratified
Additional Protocols I and II, pledging to observe the provisions of
the Treaty. (France has not yet ratified Protocol I.) Significantly,
Tlatelolco requires its parties to submit their nuclear activities to
IAEA safeguards and established an agency, OPANOL, to verify
compliance with treaty provisions.

The centerpiece of all nonproliferation agreements is the Non-
Proliferation Treaty, which entered into force in 1970, and represents
a bargain struck between the nuclear and nonnuclear weapon states.
In return for the nonnuclear weapon states' undertaking not to
transfer or receive any nuclear weapons (Articles I and II), and
submitting all their source or special fissionable material to IAEA
safeguards (Article III), the nuclear weapon states promised them 'the
fullest possible exchange' of nuclear technology (Article IV), provision
of the potential benefits from any peaceful applications of nuclear
explosions (Article V), and good faith negotiations on effective arms
control and disarmament measures (Article VI). Though the NPT
enjoys widespread support, with 115 members, several significant
holdouts remain: Argentina, Brazil, China, France, India, Israel,
Pakistan, Spain, and South Africa. Developing country governments
condemn the Treaty's invidious discrimination between the nuclear
haves and the have-nots, and can hardly be blamed for opposing the
Treaty when ratification surrenders valuable bargaining leverage in
return for few tangible benefits. Nuclear suppliers' concerns over the
dangers of sensitive nuclear technology exports have dulled their vigor
in implementing Article IV. Peaceful nuclear explosions are now considered indistinguishable from weapons tests and devoid of practical
utility, gutting Article V. And since 1974, no significant arms control
measures have harnessed US–Soviet strategic arsenals. Nevertheless,
the NPT has succeeded in reinforcing the presumption that proliferation is bad, and that its perpetrators must pay the political costs
involved in flouting international consensus.

Variety in nuclear weapons control agreements benefits all. It gives
the nonnuclear weapon states a choice of instruments through which
to express their policies, and distinguish themselves from one another.
For instance, India acceded to the LTBT while Pakistan did not;
Brazil ratified Tlatelolco while Argentina did not; Egypt signed the
NPT while Israel did not. The nuclear weapon states and other nuclear
supplier governments are glad for the range of alternatives available to
the others. Argentina and India denounce the NPT as discriminatory,
but the former has signed and promised to ratify Tlatelolco and the
latter has acceded to the LTBT. Brazil and Spain also refuse to sign
the NPT, but both have accepted safeguards over all their nuclear
activities. In each of these cases, the cause of nonproliferation has

been advanced by the halfway measures made available to nonnuclear states.

A third factor, apart from technological capabilities and official policies, intimates a country's nuclear prospects: the nature of the regime. A regime that is irredentist, outcast, ambitious, irascibly led, or otherwise unstable, may be prone to emphasize the security benefits of a nuclear policy. The more of these traits concentrated in one regime, the more security oriented a nuclear policy is likely to be. Considering Colonel Qadhafi's volatile character, his desire for a Libyan-led, trans-Sahelian state, as well as for a leading role in international Islam, and the repeated efforts to overthrow him, Libya must be considered a leading candidate among nuclear weapon-prone regimes, especially since the potential members of Greater Libya (in addition to the already compromised Chad) may hold rich uranium resources. Irredentism also is apparent in Iraq, where it contributed to the decision to attack Iran for recovery of the Shatt al-Arab, in Syria, whose October 1980 friendship treaty with the Soviets allegedly contained a secret clause providing for nuclear defense against Israel, and in Pakistan, where the Kashmir dispute still smoulders.[40] These governments also wish to enhance their regional status, as did Iran under the Shah. Regional competition also influences Argentine, Brazilian, and Indian nuclear policies, the latter being particularly interested also in maintaining strong influence over the nonaligned movement.

The nature of the Sukarno regime sometimes generated concerns over Indonesian nuclear intentions. The President's fervent nationalism led to annexation of the former Dutch colony, West Irian, and to threats against Malaysia and Singapore. His capricious nature, links with a new nuclear power (China), which was discontented with the existing nuclear power balance, and precarious political situation could have combined, it was feared, to bring about the extremely unlikely event of an Indonesian nuclear explosion. The simpler detonation of a conventional charge surrounded by radioactive pellets, though less menacing than a fission weapon, still would have set an alarming precedent.

For outcast or beleaguered states, only self-perception matters. Outside suggestions that all potential threats can easily be handled conventionally matter little, if at all, in Israel, Taiwan, and South Africa, where disappointments have schooled the governments in the irreplaceability of self-reliance. Instability spawns diverse effects. A change in government may (1) increase the prospects for a military nuclear program, as when Indira Gandhi succeeded Morarji Desai, (2) decrease them, as when Amin and Sukarno were overthrown, (3) leave them unchanged, as when Zia replaced Bhutto, or (4) leave them in doubt, as after the Iranian revolution. In addition to its influence over military postures, the nature of a regime also influences nuclear power policies.

One final set of policy determinants lies largely outside the ambit of the developing country governments. These outside constraints are tools in the hands of the nuclear supplier nations, used by them to steer importers' nuclear policies insofar as possible. The treaties and institutions described above certainly constrain nuclear policies, but since they are voluntarily accepted they cannot be considered to be external. Outside constraints, in the first instance, arise from the resistance of some governments to voluntary constraints. Due to their refusal to bend their nuclear policies to the desires of the nuclear suppliers, India, Pakistan, and South Africa have been deprived of North American nuclear assistance. Many more governments have been buffeted by outsiders' dissatisfaction with their activities, as the Americans pressed heavily for the South Koreans to abandon reprocessing plans and the Taiwanese to dismantle a reprocessing laboratory. These examples show that external pressure may be bilaterally applied. The North Americans are best known for, but not alone in, this approach. The Australians suspended all uranium exports during the Fox Commission inquiry into the advisability of exporting uranium from the Ranger mines, from health as well as nonproliferation perspectives. The French reneged on their commitment to supply a reprocessing plant to Pakistan, and tried to dissuade the Iraqis from demanding weapons-grade enriched uranium.

Most governments, however, are extremely reluctant to apply pressure bilaterally. They do not wish to compromise their commercial reputations as reliable suppliers or otherwise spoil relations with developing nations, and fear the futility of bilateral constraints, which usually can be circumvented simply by resorting to the competition. Multilateral constraints, though more difficult to achieve, are consequently preferable. The most extensive multilateral effort to constrain nuclear exports and so influence others' nuclear policies took place in London, at the Nuclear Suppliers' Group, which was convened in the aftermath of the Indian nuclear test in order to regularize export policies. Specifically, the seven (later expanded to fifteen) members of the group sought to isolate commercial competition for nuclear exports from nonproliferation requirements, so that competitors would not try to outbid each other in the sensitivity of technologies offered, or the laxity of safeguards required.[41] To this end, in February 1978, the NSG members separately submitted a set of nuclear export guidelines, which included a list of specific items whose export would trigger the application of safeguards.[42] Disagreements were to be resolved through consultation, not preemption.

The success of the NSG guidelines has been mixed. Both Argentina and Pakistan have successfully evaded them. Without doubt, though, they incurred the unbridled animus of the nonaligned governments, who universally condemned them as an arrogant extension of the already discriminatory nonproliferation regime. A typical sentiment was expressed by AEOI President Etemad, when he remarked that no country or group of countries had 'a right to dictate nuclear policy to another'.[43] The vociferous response to the guidelines plainly revealed their strong political impact. From the suppliers' perspective, they marked an important step toward cooperation and recognition that nuclear power and nuclear weapons were inextricable. During the course of the NSG discussions, President Valery Giscard d'Estaing and Chancellor Helmut Schmidt pledged to abstain, at least for a time, from further agreements to sell reprocessing facilities. The NSG members paid for their achievement by adding to the developing world's arsenal of discrimination allegations.

Voluntary submission to constraints is refused when external pressures are less hurtful than the political gains to be made from flaunting independence in the face of adversity. Governments' decisions hinge upon their devotion to nonaligned solidarity compared to their desire for smooth nuclear development and strong security guarantees. If external sanctions are strengthened or eased, then some governments might slip from the intransigent to the tractable categories or vice versa.

CONCLUSION

No clear result, which could be called the 'net security impact' of the balance between the pros and cons of nuclear power, emerges from this analysis. Each government trades off some benefits for other costs. In choosing to acquire nuclear weapons, each must weigh the prestige and regional influence gained against the possible loss of Western aid or security guarantees. In choosing to build a commercial reactor, each must compare the relief from future vulnerability to OPEC to the possible domestic unrest provoked by the diversion of so many resources from present consumption to capital investment. Without access to government councils, an assessment of official policy, technological capabilities, political and energy factors, provide the best clue to how much security objectives have influenced nuclear power policies. Take, for example, Iran's decision to build 23,000 MWe of nuclear power in twenty years. The Shah consistently and unambiguously rejected nuclear weapons acquisition, but mistrust of his objectives sprang from his immense military budgets, his affinity for advanced technology for its own sake, and suspicions that his vision of the future included a lengthened reach from the Peacock Throne.[44]

Some feared that the attraction of earning a seat among the great nuclear powers would prove irresistible to the monarch. This fear might have been justified. Further, it seemed blatantly illogical to spend $30 billion or more for twenty-odd reactors, based only on the vaguest electrical demand projections, in a country possessing immense natural gas reserves. It seemed sinful that, due to a lack of investment, over half of the annual natural gas production was wasted, simply flared at the well-head. Such apparent irrationality could be explained by positing that economics were irrelevant to the nuclear program's real objectives, one of which could have been nuclear weapons.

Nevertheless, such speculation cannot explain either the Iranian selection of the light water reactor cycle (the one least conducive to weapon uses), the lack of movement toward obtaining weapons-grade material-producing technologies such as uranium enrichment and plutonium reprocessing, or why the Shah would spend so much money for such a gargantuan cover for a weapons program (especially since its very size was a major source of suspicion). The Shah may have wanted to earn a seat among the great powers, not through nuclear weapons, but rather through his nuclear power program, through the prestige and independence he could gain from its grand scale. He may also have desired to acquire nuclear weapons, but these desires did not shape his nuclear power program.

By contrast, the Argentine National Atomic Energy Commission (CNEA) selected the heavy water technology, which is more suitable to manufacture of a plutonium fission device, and constructed an indigenous and unsafeguarded, if small, plant to extract plutonium from spent fuel. The Argentinians were better able than the Iranians to manufacture nuclear weapons. Repeated government affirmations of peaceful intent must be weighed against refusals to accede not only to the NPT, but also to the Treaty of Tlatelolco, which was supported even by its regional rival and fellow-NPT opponent, Brazil. Although it is far from certain that the Argentine government has ever intended to develop nuclear weapons, its selection of the most militarily applicable technologies and resistance to international safeguards has left the nuclear weapons option open. Even if the option is never pursued, its existence affords Argentina a useful lever to win concessions from the industrialized nations.

Since the division of security objectives into military, political, and energy components is artificial, it is not surprising that potentially important objectives fit none precisely. One such aim, falling between the military and political categories, is to foster *suspicions* that a government may acquire nuclear weapons, without actually intending to do so. If a government does not detonate a nuclear device and avers its benign intentions, then it cannot be termed an international villain or a nuclear weapon state. But so long as it encourages suspicions that it *might* acquire nuclear weapons, it enjoys two advantages. First, it obtains a limited regional deterrent as well as added prestige. In deciding whether to acquire their own nuclear weapons or to pursue an extremely hostile foreign policy, possible adversaries have to weigh the possibilities that a government might respond by using civil nuclear technology for military uses. Suspicions can translate into increased regard for the political will and military proclivities of a regime, thereby giving governments in the countries studied a freer hand and more sway in regional affairs.

Second, ambiguity of nuclear objectives can be used against the suppliers of nuclear technology and materials who wish to prevent the development of nuclear weapons by restricting exports of dangerous technologies. So long as refusals persisted by Argentina to ratify the Treaty of Tlatelolco, Iran to forswear reprocessing, and Indonesia to ratify the NPT, such issues could be used as levers to pry loose exports which the advanced countries otherwise might refuse to sell them. This strategy cost little because, among the nonnuclear weapon states, there is little discrimination by the nuclear suppliers. The NPT is supposed to reward nonnuclear weapon state adherents by giving them preferential treatment in obtaining peaceful nuclear technology and the benefits from peaceful nuclear explosions, while the nuclear weapon state parties trammelled the arms race.

Naturally, IAEA members which have not acceded to the NPT insist that access to nuclear technology was promised by Agency membership and cannot be prejudiced by subsequent treaties. As time passed, however, certain technologies were deemed too sensitive to transfer, even to NPT parties, and peaceful nuclear explosions were deemed useless and indistinguishable from military nuclear explosions. Meanwhile, US and Soviet nuclear arsenals continued to grow.

Ironically, in order to dissuade the most worrisome non-NPT parties from developing nuclear weapons (and in order to make profits), the governments and vendors in the developed countries have sought to tie them into alternative nonproliferation obligations through continued or expanded nuclear exports, under international safeguards. For example, after India detonated a nuclear device, American officials continued to countenance sales of enriched uranium there (albeit with delays), believing that to cut off nuclear co-operation with India entirely (as Canada had) would eliminate all remaining leverage which the United States might otherwise continue to exercise in India. As the years elapsed, the lack of further Indian tests was cited to illustrate the wisdom of this policy.

In short, nonnuclear weapon states of any color can reap all available benefits from nuclear trade. Those party to the NPT have had to submit *all* nuclear facilities to IAEA safeguards, in return for broken promises. Nonparties have had to accept safeguards only on facilities imported from parties to the Treaty, retaining the option to eschew safeguards for indigenous facilities or for materials received from other nonparties. For those particularly uncooperative governments which have insisted on maintaining (Argentina) or exercising (India) this option, some nuclear suppliers argue that it is safer to sell them a safeguarded facility – even one as sensitive as a heavy water plant – than to provoke them to build their own, unsafeguarded plants. The suspicious behavior of such countries adds still more leverage for use against the developed countries. These strategic advantages can be attained regardless of a government's *actual* military objectives. Sincerity, if it exists, cannot govern future actions. At any moment, any indication of a possible nuclear weapon program can furnish a security advantage.

NOTES

1 'Remarks by President Carter on nuclear power policy, April 7, 1977', *Presidential Documents: Jimmy Carter 1977*, vol. 13, April 18, 1977; and United States, *Nuclear Non-Proliferation Act of 1978*, Pub. L. 95–242, March 10, 1978.
2 See Joseph A. Yager (ed.), *Nonproliferation and U.S. Foreign Policy* (Washington: The Brookings Institution, 1980).
3 P. F. Wilmot, 'The future of Africa: revolution, retrogression or inertia', Ahmadu Bello University (Nigeria) Public Lecture, April 23, 1981, p. 12. Ali Mazrui suggested that 'the "vaccination" of horizontal nuclear proliferation might be needed to cure the world' of vertical proliferation, adding more warheads to the enormous nuclear strike forces already in existence, in 'Third World security: a cultural perspective', *International Affairs*, vol. 57, no. 1 (Winter 1980–81), p. 18.
4 Pronuclear sentiments among the 'concerned' Indian public constrains government policy, according to F. T. J. Bray and M. L. Moodie, 'Nuclear politics in India', *Survival* (May–June 1977), pp. 112–13. Political party (including Janata) support for nuclear weapons is discussed in O. Marwah, 'India's nuclear and space programs: intent and policy', *International Security*, vol. 1, no. 4 (Spring 1977), p. 97, and *The Overseas Hindustan Times*, May 14, 1981, p. 4.
5 Betts, in Yager (ed.), *U.S. Nonproliferation*, op. cit., p. 160.
6 Many believe that a nuclear weapon test caused the double flash recorded in the South Atlantic by a US Vela satellite on September 22, 1979, and that South Africans were responsible. (Israel is suspected of complicity.) Minister for Mines and Power, Alhaji Mohammed Hassan, announced that Nigeria would embark upon a nuclear program, left undefined, partly in response to the nuclear trade undertaken by non-NPT members, presumably South Africa. *The Times*, July 2, 1981.
7 Quoted in *Chicago Tribune*, February 1, 1978, pp. 9, 10.
8 Richter, quoted in *New York Times*, March 25, 1951, p. 10; see Brazilian charges

of Argentine nuclear weapons program, *Le Monde* (Paris), July 8, 1967; Quihillalt denial appeared in *Christian Science Monitor* (London edn), August 5, 1967.

9   *Nucleonics Week*, June 26, 1975, p. 9.

10 J. M. Ruda, 'La posicion argentina en cuanto al Tratado sobre la No Proliferacion de las Armas Nucleares', trans. N. Gall, *Estrategia* (September 1970–February 1971), p. 79.

11 'Treaty on the Non-Proliferation of Nuclear Weapons,' in United States, Arms Control and Disarmament Agency, *Arms Control and Disarmament Agreements*, 1977 edn (Washington, D.C.: US ACDA, 1977), Articles IV, V, and VI.

12 'Treaty for the prohibition of Nuclear Weapons in Latin America', and 'A Proclamation by the President of the United States of America, concerning Additional Protocol II to the Treaty for the Prohibition of Nuclear Weapons in Latin America, June 11, 1971', in United States, *Arms Control Agreements*, op. cit., Article V and Section II, respectively.

13 For discussion of this agreement and its implications, see R. Gillette, 'India and Argentina: developing a nuclear affinity', *Science*, vol. 184, June 28, 1974, pp. 1351–53.

14 *Nucleonics Week*, September 12, 1976, p. 2.

15 For details of agreement see FBIS, *Worldwide Report: Nuclear Development and Proliferation*, no. 48, June 13, 1980, pp. 31–6.

16 *Daily Telegraph* (London), July 29, 1965, p. 23.

17 *Christian Science Monitor*, May 13, 1966.

18 Suharto's ability to relinquish this symbol without great political cost might have stemmed not only from the wealth of alternative energy resources in Indonesia, but also from the tendency of the Indonesia political elites to follow any presidential policy or its reversal, so long as resources are supplied for their personal wants. More than in most political systems, the personal beliefs of the chief executive in Indonesia leave their stamp on policy outcomes large and small. K. K. Jackson, 'Bureaucratic polity: a theoretical framework for the analysis of power and communications in Indonesia', in K. D. Jackson and L. W. Pye (eds), *Political Power and Communications in Indonesia* (Berkeley: University of California Press, 1978), p. 17.

19 *Kayhan International*, March 8, 1974, p. 4.

20 Quoted in Karanjia, *Monarch*, pp. 242–3.

21 R. Tomar, 'The Indian nuclear power program: myths and mirages', *Asian Survey*, vol. 20, no. 5 (May 1980), pp. 527–9.

22 *Clarin* (Buenos Aires), trans. Foreign Broadcast Information Service, February 18, 1979, pp. 4–5.

23 For details on the military nuclear capabilities in various threshold nuclear states, see Yager (ed.), *U.S. Nonproliferation, passim*.

24 Quoted in *Chicago Tribune*, February 1, 1978, p. 9.

25 J. A. Sabato, quoted in *The Washington Star*, October 6, 1975.

26 E. S. Milenky, *Argentina's Foreign Policies* (Boulder, Co.: Westview Press, 1978), p. 36.

27 CNEA, *Centro Atomica Ezeiza* (np., 1967); and V. Johnson and C. Astiz, 'Latin America', in F. C. Williams and D. A. Deese (eds), *Nuclear Nonproliferation: The Spent Fuel Problem* (New York: Pergammon Press, 1979).

28 *Financial Times*, February 9, 1979, p. 4.

29 K. K. Pathak, *Nuclear Policy of India* (New Delhi: Gitanjali Prakashan, 1980), pp. 145–9.

30 *Statesman Weekly*, New Delhi, January 3, 1981.

31 For the relationship between actual and boasted power, see Jackson, 'Bureaucratic Polity', op. cit., p. 39. He argues that in Indonesian culture, power is best shown by how little one has to exercise it openly. By this logic, hyperbole is cheap, and its use

with reference to a nuclear weapons program could have been interpreted as a confession of Indonesians' inability to manufacture them. A genuine threshold nuclear power would not need to resort to such brash threats.

32   Reported in *Sunday Times* (London), June 30, 1974.

33   *New York Times*, April 2, 1975, p. 2.

34   *Times of India*, April 26, 1981, as quoted in *The Observer* (London), May 3, 1981.

35   A. G. Noorani, 'Indo–U.S. nuclear relations', *Asian Survey*, vol. 21, no. 4 (April 1981), p. 400.

36   For the request to China, see M. Heikal, *The Road to Ramadan* (London: Collins, 1975), pp. 76–7.

37   *Iraq* (Iraqi Cultural Centre, London) (June/July 1981), p. 1.

38   W. Epstein, *The Last Chance* (New York: The Free Press, 1976), pp. 207–20.

39   Article 28 stipulates that the Treaty shall enter into force once all Latin American Republics have ratified it, all states responsible for territories within the Treaty's jurisdiction have ratified Additional Protocol I, and all nuclear weapons powers have ratified Article II. Cuba is unlikely to sign, but all the signatories except Argentina, Brazil, and Chile have waived the entry into force requirements.

40   *The Observer* (London), November 9, 1980, reported that a secret clause in their October 1980 Treaty of Friendship obliged the Soviet Union to take all necessary steps, including the threat of nuclear reprisal, to prevent Israel from using atomic weapons against Syria.

41   The original members were Canada, France, Great Britain, Japan, the Soviet Union, United States, and West Germany. They were later joined by Belgium, Czechoslovakia, East Germany, Italy, the Netherlands, Poland, Sweden, and Switzerland.

42   Reprinted in *Survival* (March–April 1978), pp. 85–7.

43   US Department of State, *Telegram Tehran 1232*, February 7, 1977, para. 3.

44   F. Halliday, *Iran: Dictatorship and Development* (Harmondsworth, Middlesex: Penguin, 1979), p. 268. For the Shah's desire to acquire the most advanced technologies, see Pahlavi, *Mission*, op. cit., p. 137.

8   Economic Objectives

Ever since the 1950s, nuclear energy has been promoted as a potentially important element of future economic development for the poorer countries, one so complex that eventual exploitation requires immediate study. With the help of US Atoms for Peace grant assistance, many interested governments purchased small research reactors for experimentation with radioisotopes and training scientists. From these beginnings, a nuclear power program may emerge. In deciding whether to proceed beyond the first stages, governments gauge the economic costs and benefits nuclear power may entail. This chapter will evaluate the influence of that calculation over nuclear policy, first, by studying the relationship between nuclear power and the central economic objectives and, second, by examining how this relationship applied to the independent, dependent, and nonnuclear developing countries.

GOVERNMENT BY NECESSITY

According to Aristotle, government is created to preserve mere life, but is maintained to promote the good life.[1] Once the stability prerequisite to economic development is attained, governments can devote their attention to 'the good life', often identified with income growth and industrialization. There are, however, other economic goals, which call for radically different policies.[2] Savings and investment help increase future consumption, but at the expense of present consumption. Efforts to attain independence impose the costs of autarky. Projects that maximize production are not always those which most relieve destitution, provide employment, or equalize consumption.

Resolution of these conflicts is a governmental task, for two major reasons. First, the government arbitrates between opposing interests. Since conflict between them willy-nilly will shape the politics and stability of any regime, no government can afford to relegate central economic decisions to the laws of nature or of the marketplace. Second, even if the government wished to abdicate in favor of the rule of nature or of the marketplace, that rule left to itself would produce economic disaster. Prices may be a poor guide to resource allocation in the energy sector due to international market distortions and the lack of integrated national markets. Even if purely free markets existed, however, the energy sector would be a natural exception to them. The development of new energy technologies takes so long and is so expensive that even the governments of the industrialized world feel obliged to underwrite research and development, since the scale of investment is usually high while returns seems uncertain and remote. Government interference in the energy market is further justified by the reliance of every other economic sector upon a secure supply of energy.

Other reasons compel governments to intervene in resource allocation more in the developing world. One is uncertainty, which confounds investors everywhere but plagues those in developing countries especially. Entrepreneurs can be frozen into inaction by the fear that price increases may cancel out the returns on investment, especially when inflation reaches triple-digit annual levels, as it has in the last two decades in many developing countries.[3] The many governments which depend heavily upon export revenues are vulnerable to the additional uncertainty of fluctuating foreign demand. Reduced demand can cripple economies kept solvent by the export of one or two items, and recessions in the developed economies are often amplified there by the lack of domestic demand or alternative outlets for their productive abilities and resources.

Second, externalities compromise the effectiveness of market-based resource allocations. Externalities are effects of an activity not accounted for in its price. Pollution is the most common example. People suffer from it, but this suffering is seldom quantified and incorporated into the calculation of the costs of building a factory, nor is it often compensated. Tariffs and currency overvaluation are used to protect local industries from succumbing to the competition of better and cheaper foreign imports; both measures impose upon domestic consumers costs which often are not taken into account. Currency overvaluation hinders foreign investment, while pricing the developing country's exports out of the world market. External costs can be compensated by external or 'spillover' benefits, such as the increased productivity of workers trained for one job who later apply their improved skills to other projects. But it is unlikely that external costs and benefits will balance precisely, especially where governments systematically tend to favor certain interests.

Certain externalities are unique to nuclear power. Decommissioning costs cannot easily be internalized. An old thermal plant can be abandoned safely, to decay in retirement. Its parts can be reused or melted down. A nuclear power station, however, houses radioactivity and cannot simply be left untended. Instead, the plant must be 'decommissioned', sealed off, and stripped of radioactive components, which then must be safely disposed. The costs of this process remain speculative, since no commercial-scale nuclear station has yet been decommissioned. Nuclear proponents argue that decommissioning will not add more than 10 percent of original costs in constant dollars, which reduces to less than 1 percent when discounted at 5 percent over fifty years.[4] Nuclear critics, on the other hand, charge that decommissioning costs could be so significant as to vitiate any potential economic benefit of nuclear power. Insurance costs may also be at least partly externalized, because nuclear power is an actuarial anomaly. Normally, insurance costs are included in any project's cost–benefit analysis. The premiums, conditions, and benefits of an insurance policy are based upon the product of the probabilities that accidents of various severities will occur. Attempts to draft insurance policies for nuclear power are enormously complicated, because although the probabilities of a major accident may be judged infinitesimally small (on the order of 10^{-6} per year, according to the Rasmussen report on nuclear safety), the damage which could arise from a major accident could be incalculably large, far beyond the capacity of most if not all underwriters to indemnify. In this, nuclear accidents resemble the 'acts of God' from which insurers traditionally exempt themselves. Since no clear limits to liability exist, insurance tends to be either ignored or artificially limited. The US Price–Anderson Act limited liability for a nuclear accident to $560 million. The IAEA recommended a minimum of $5 million insurance for a nuclear power station. In Taiwan, the $5 million recommended minimum has been adopted as a maximum. The effect of artificial constraints on liability is to enhance the economic attraction of nuclear power at the expense of increasing externalized risk to the consumer.

Coping with uncertainty and externalities is the task of government. Good state planning clarifies tradeoffs, quantifies them as much as possible, and then helps arbitrate between them. According to UN Industrial Development Organization (UNIDO) guidelines, the central planning organization should fill the gap between commercial profitability and national gains.[5] Ironically, the political process resists overtly choosing between the values which planners make explicit. Stark exposure of economic choices highlights conflicts of interest, while showing which parties stand to gain or to lose most from any given project. Political stability, however, demands that conflict be smoothed over and, if possible, obscured.

Its inherent bias does not render planning useless; indeed, the very attempt to inject some rationality to the policy process is likely at least to seep into the thought processes of project formulators, and may influence which projects the politicians choose to back. Gradually, the presumption may shift against extremely wasteful projects, or those directed only at gaining prestige. This process has already become apparent in many quarters, and has sometimes cut against nuclear projects, though not always successfully, in countries such as Argentina, Brazil, Indonesia, and Iran.

One of the planner's first tasks must be to assess the relationship of the energy sector to economic growth in his country. In general, the larger the share of national income which is devoted to the procurement of useable energy, the greater will be the toll extracted from economic growth whenever energy prices rise disproportionately to other goods. Also, to the extent that energy can be replaced by other inputs in economically productive activities, the economy will be partly shielded from the harms of energy price increases. In the industrialized countries, energy-related investment comprises a small share of GNP (less than 2 percent in the United States) and apparently can be substituted at no great loss. The ratio of energy use to GNP has been declining in the industrialized nations since 1973. The developing countries' smaller economies are more dependent upon energy. Since there is less of a cushion in infrastructure and alternative investment patterns there, these economies cannot easily rearrange consumption to minimize the impact of oil price increases, which fall heavily upon scarce foreign exchange and capital reserves. Since consumption is already minimal, these expenditures come more at the expense of savings and investment than in the industrialized nations. In unfortunate contrast to the North, the share of energy in GNP has been rising in the South, as Table 8.1 reveals. Planners can help reduce the taxing effects of energy price hikes by recommending the development of an industrial strategy weighted toward low energy intensive activities.[6]

The issue at hand is not whether nuclear power in fact can fulfill the economic objectives of Third World governments, but whether they think it can, and why. The prospects for nuclear power development are apparent in the overall development plans common to developing countries. If it is not given top priority, nuclear power stands little chance, for only an intensive effort to marshal the technical and financial infrastructure required can overcome the economic deficiencies in most developing countries.

NUCLEAR ENERGY AND ECONOMIC OBJECTIVES

Were governments to abjure intervention in the marketplace, they probably would still feel compelled to regulate the energy sector, for

Table 8.1 *Economic Growth and Commercial Energy Consumption, 1970–78, Selected Countries (percent per year)*

	Economic growth rates		Growth in commercial energy consumption		Change in energy intensity	
	1970–73	1973–78	1970–73	1973–78	1970–73	1973–78
Algeria	5.3	5.5	14.4	11.3	9.1	5.8
Brazil	12.8	7.2	12.8	7.9	0.0	0.7
Colombia	7.2	5.6	5.4	3.9	−1.8	−1.7
Egypt	4.4	11.1	5.6	11.9	1.2	0.8
India	1.8	4.4	4.3	4.8	2.5	0.4
Indonesia	8.7	6.7	13.4	18.4	4.7	11.7
Jamaica	4.6	−2.5	4.5	−0.6	−0.1	1.9
Kenya	5.5	4.8	8.8	3.7	3.3	−1.1
Korea	9.7	10.3	5.8	9.1	−3.9	−1.2
Mexico	6.0	3.5	7.3	6.9	1.3	3.4
Nigeria	7.0	7.0	20.1	8.5	12.9	1.5
Philippines	6.8	6.1	3.9	4.5	−2.9	−1.6
Portugal	8.9	1.6	15.5	2.3	6.6	0.7
Thailand	6.9	7.3	11.8	5.1	4.9	−1.8
Turkey	7.7	6.1	11.5	7.7	3.8	1.6
Venezuela	4.4	7.7	7.1	4.0	2.7	−3.7
All developing countries[a]	6.7	5.3	8.0	7.3	1.3	2.0
All industrial countries[b]	5.4	2.9	4.1	1.5	−1.3	−1.4
World	5.6	3.4	4.8	2.5	−0.8	−0.9

[a]Including OPEC members, China, and other Asian centrally planned economies.
[b]Western and Eastern Europe (including the Soviet Union), the United States, Canada, Japan, Australia, and New Zealand.
Source: J. Dunkerley, W. Ramsay, L. Gordon, and E. Cecelski, *Energy Strategies for Developing Nations* (Baltimore, Md.: Johns Hopkins University Press, 1981), pp. 89–90 published for Resources for the Future Inc.).

several reasons. Energy is essential to all economic activities, and governments cannot remain indifferent to the security implications of major arterial networks in society. In laying this network and requiring their citizens to use it, governments increase their domestic political control. Through network maintenance and protection, governments ensure the citizen's dependence as well as their own insulation from foreign pressures. Moreover, oil companies, electrical utilities, and other segments in the energy generation and transmission market are naturally monopolistic, their marginal costs falling with each increased unit of production. Its security implications and natural market characteristics often combine to encourage the nationalization of the energy sector or of several of its constituent parts. Once nationalized, the increased scale of demand enables the government to promote scale economics in supply. The 'lumpy' supply inherent in increased unit scale requires that, similarly, investment funds be acquired in large amounts at irregular intervals, and that supply and demand be carefully coordinated, as shown in Figure 8.1.

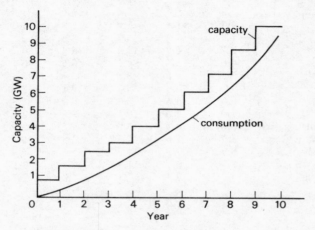

Figure 8.1   *Energy consumption and capacity growth*

Ultimately, the role of nuclear power depends upon the role assigned to electricity in a country. Electrification has both political and economic dimensions. Most developing country governments proceed from the premise that electricity supplies are in severe shortage. The causes vary, but chief among them is population-driven increased demand. Not only has population in the South mushroomed since 1945, with average annual growth rates still exceeding 2 percent, or more than double the industrialized country rate, but it has become more and more concentrated as people have flooded into cities. The middle-income developing countries (those with over $360 GNP per capita) have been particularly hard hit, with the average urban share increasing from 37 to 51 percent since 1960.[7] As per capita national income has increased, so has demand for electricity, with uses extending beyond illumination, communication, and refrigeration, to less essential items such as air conditioning and home appliances. From 1960 to 1977, electricity production in the developing world increased from 202.2 billion kWh to 641.3 kWh, outpacing electrical production growth rates in the developed world over that period.[8]

Multiplying demand has often caused severe shortages. In Argentina, an influx of four to five million rural immigrants to Buenos Aires under the first Peron regime (1946–55) pushed electrical demand far ahead of available supply. Governments did not always help. Peron aggravated the problem by enforcing an 'electrical diet' for the capital area, which barred investment in old power plants and the construction of new ones.[9] Complaints of electrical shortages have continued to the present.[10] On Java, where hotel guests find contingency candles in their nightstands, estimated demand for electricity which was suppressed by inadequate supply by 1973 reached over 1,160 MW, 80 percent more than installed capacity.[11] In Nigeria, the performance of the National Electrical Power Authority has led to suggestions that its initials should be interpreted to mean 'Never Electric Power Again'. The politically destabilizing power outages in Tehran in 1976 and 1977 have been noted. While governments seek to extend political control by extending reliance on public electricity grids, they concomitantly incur the blame for system failures, which can have harsh political consequences.

Many developing country governments aim to extend electrical services to all citizens, or at least to all citizens who live near urban centers. Others wish to increase the value of their traditional exports by processing them, for example, by smelting ore into metal.

Expanded electricity supplies are also desired to promote industrial productivity (thus increasing economic output) and to facilitate enjoyment of the goods and services produced (thus stimulating further demand).[12] Erratic energy supply not only inconveniences individuals, but also discourages industrial demand for state-supplied electricity. Businessmen who rely upon continuous current often choose to buy and operate their own generators. One half or more of the installed electrical generating capacity in many developing countries is 'captive'.[13]

Greater scale economies could lower the costs of electricity production for all, if firms were to purchase state utility services rather than generate their own power from these relatively small units with high unit capital and operating costs. Reduced captive power would also cut consumption of diesel oil, while conserving the foreign exchange used to import generators for other productive investments. As state utilities gain more customers, more revenues flow into state coffers, further encouraging some governments to eliminate captive power plants. Centralization of political and economic power increases as outsiders are absorbed into or eliminated from a nationalized energy sector.

Nuclear power provides large increases in electrical supplies, obviating the need for captive plants, which may be kept as emergency generators. Throughout the 1960s, it was still seen, by and large, as too complex, unproved, large scale, and uncompetitive with fossil fuels to warrant commercial investment in any developing countries except India, Pakistan, and Argentina. By the mid-1970s, though, circumstances had changed. In many developing countries demand was increasing to the level at which nuclear power could at least be considered, while the complexity began to appear more a challenge and less a threat. Also by this time, the technology no longer seemed unproved; over 61,000 MWe of nuclear power had entered commercial operation worldwide, including over 1,000 MWe in India, Argentina, and Pakistan. Perhaps most importantly, oil price increases catalyzed the quest for alternative energy resource development. Coal was seen as a dirty, bulky fuel of the past, obtained from an obsolete and decaying industry. Among all alternatives, nuclear technology appeared most ready to fill the breach. So, encouraged by optimistic IAEA studies and eager nuclear reactor suppliers, many developing country governments began serious planning toward the introduction of nuclear power.

BOON OR BURDEN?

Before turning to the method of choice, it is useful to identify factors common to many developing countries which tend to favor or oppose investment in nuclear energy. In addition to the traditional land, labor, and capital categories, exports will be discussed separately, since nuclear power programs in developing countries powerfully influence trade and current account balances.

On the credit side, nuclear power can help labor in many ways. Electrical lights and appliances improve conditions at home and in the workplace. The introduction of electrical manufacturing and industrial processes can increase productivity and accelerate growth. The nuclear power project itself employs and educates. Over a thousand domestic workers (depending on how much skilled foreign labor must be imported) may participate in the construction of a station, and over a hundred may be needed at a time to operate and maintain it. In the course of the project, technicians in many fields receive on-the-job training, which can then be applied to other projects. The first group, trained by foreigners, can take over the task of training their countrymen, without language or cultural barriers. Domestic talent may better understand the special needs of the country and the modifications which may increase the applicability of foreign technology to local circumstances. In Argentina, for example, professional personnel at the CNEA rose from a few dozen in 1956 to around one thousand in the early 1970s. Once construction began, hundreds of skilled Argentinians could be found on the Atucha site. Every year, the CNEA sent professionals abroad for training. Building upon a fairly sophisticated educational base (though one compromised by political intervention), the CNEA trained and hired hundreds of skilled technicians and workers. 'Throughout its history', wrote the former director of technology, 'CNEA gave first priority to the training of scientific and technical personnel.'[14] Training was not confined to areas applicable only to nuclear power, which enabled personnel trained by the Commission to move out and work in universities, industries, other governmental research institutions, and hospitals.

On the debit side, the serious skilled labor shortages afflicting most developing countries can seriously reduce the value of nuclear power. First, nuclear power reactors require a high degree of quality control, so some workers must be trained to perform difficult tasks extremely well. Of course, many tasks in building a nuclear station are simple and common to any construction project. But insofar as a skill particular to nuclear projects spills over to other sectors, the per capita investment in cultivating it will greatly exceed that for many other occupations. Often, a greater investment in training is required because of the much lower initial educational base; illiteracy, tardiness, absenteeism, and carelessness must be conquered before more sophisticated skills can be taught. Where skilled labor shortages are severe or where project managers lack patience, there is no choice but to import labor. Payments to these interlopers subvert the spill-over process. At project's end they take whatever they have learnt on the job back home. Hiring foreigners also suppresses aggregate demand, by draining the economy of that portion of their salaries which they send to their families abroad. Moreover, educated domestic workers deeply resent working next to foreigners with similar qualifications but much higher salaries. The larger the foreign contingent, the greater the irritation.

The quantity and quality of labor must be traded off against each other. Suppose a $15,000 outlay can be used to train either fifteen typists, or seven carpenters, or one engineer. Deciding the best way to spend the $15,000 requires evaluation of economic needs, and of the opportunity cost of whatever decision is taken. The opportunity cost of training one engineer is equivalent to the training of seven carpenters, or fifteen typists, or three carpenters and eight typists combined. The relative capital or labor intensity of different projects imposes additional opportunity costs. If a $1,000 project, *A*, is comprised of an $800 machine and a $200 salary for its operator, while a similar project, *B*, requires a $200 machine and four workers earning $200 salaries, then *B* is more labor-intensive than *A*, providing four times as many jobs per dollar invested. Nuclear investments are extremely capital intensive, and in the labor surplus economies of the South, labor intensive activities could immediately offer more benefit to more people. The long-term effects are less clear. But in the short term, both the high level of training required for nuclear technicians and the few domestic jobs provided per dollar invested (reduced further by foreign labor) impose a heavy opportunity cost, and reduce any employment benefit from nuclear projects.

Turning from labor to land, the benefits from nuclear power depend upon a country's natural resource base, and the costs of exploiting uranium versus other energy resources. As with labor, nuclear power involves land input as well as output. The net land benefit of a nuclear reactor depends not only upon the price and availability of uranium and other natural inputs (limestone for cement, iron ore for steel), but also upon the untapped resource outputs (such as electrolytically-produced metals) which electricity can prise from nature. Finally, one must attempt to weigh the cost imposed by nuclear power upon the environment, in order to reduce externalities in decision-making calculations.

The influence of capital is bound to be large, since it represents 70 percent of the cost of a nuclear power station. The heavy initial investment must be compensated by low operating and maintenance costs or by high lifetime return on investment, for the project to be profitable. The returns should account for the industrial infrastructure added which will benefit other projects. Transmission and distribution (T&D) network costs will vary between projects, but can comprise two-thirds of the cost of delivered electricity. The longer the transmission lines, the more electricity is lost in transit, due to resistance in the lines and imperfect insulation. Generally, line losses hover around 10 percent, representing an equivalent cost increase, though they may be much higher or lower. The conservation-minded Taiwanese have reduced line losses to around 6.6 percent.[15] Hydroelectric dams, far removed from consumption centers, require longer transmission lines than nuclear stations; small generators, in a decentralized system where each unit serves a smaller area, require shorter ones. The maximum reduction of transmission lines from nuclear stations equals the minimum distance permitted by political prudence and safety regulations between people and a potential radiation hazard.

As with land and labor, any use of capital entails an opportunity cost, equal to the present discounted value of other projects which might be completed with the investment funds in question. Additional complications arise through capital market imperfections. If these did not exist, and future world demand and interest rates could be predicted with confidence, then external indebtedness and foreign exchange reserves could be left for the market to determine. Imperfections, however, are endemic, and there is no sign that governments will cease to interfere with interest and foreign exchange rates. Consequently, an independent judgment concerning acceptable levels of foreign indebtedness and foreign exchange reserves must be included in major project appraisals. Much of the cost of a nuclear project must be dispensed in foreign exchange, because many key components – mechanical and electrical equipment and plant engineering, which together comprise roughly 68 percent of the capital cost – can at present only be manufactured by the industrialized nations and India.[16] This near monopoly may soon be broken by the emergence of heavy nuclear components industries in Korea and Brazil, where the Nuclep plant was built in such optimistic days that now its managers are strapped to find enough orders for the one or two reactor cores it can manufacture annually.

The pursuit of independence from foreign suppliers merely pushes the capital and foreign exchange costs even higher in the short to medium term, because outsiders are needed for the construction of the nuclear support industry. According to a Kraftwerk Union official, a minimum of eight nuclear power plants must be built in a given time period to justify over 80 percent domestic participation in equipment manufacture.[17] One way or the other, governments will need to borrow hundreds of millions of dollars to build nuclear power plants or their own nuclear industries, and to pay back principal and interest in precious foreign currencies. Table 2.3 illustrated the heavy indebtedness which already restricts many governments' freedom of economic choice. Overall, its capital expense significantly deters investment in nuclear power, and without other major economic or security advantages, most developing countries will refrain from taking up the option.

Like the foreign import component, export considerations cannot easily be internalized. The notion that someday a country might become an exporter of nuclear technology to other Third World nations is often entertained with relish, but in the first instance the cost of adopting the nuclear option is so great that it could not even be considered unless nuclear power is deemed essential to domestic energy policy. Nuclear technology export possibilities may not arise for decades, if demand for it survives at all, and even if it does the competition from the nuclear suppliers could well prove unbeatable except for small-scale assistance. To bank on nuclear exports would be foolhardy at a moment when even the nuclear exporters of Europe and America are floundering. As a nuclear program matures, though, export opportunities may present themselves, offering prestige and perhaps a hope for eventual profit to the donor. Under the aegis of the IAEA, personnel from such countries as Argentina and India have already engaged in technical assistance to other developing countries.

Bilateral assistance within the developing world has also begun. In 1980 Argentina and Brazil agreed to exchange nuclear technology, and earlier each had signed agreements for peaceful nuclear cooperation with most Latin American nations. India's nuclear clients have ranged from Argentina to Libya to Iran. In the past these pacts have existed more in word than deed, but this is changing. By 1981, Brazil had won eight major civil engineering contracts in Latin America, Africa, and the Middle East, and service export earnings for the year were expected to reach $2 billion.[18] The Argentine CNEA built a research reactor in Peru, offered to do the same in Uruguay, and expressed the desire eventually to offer reactors and fuel cycle facilities for export.[19] Brazil obtained some relief for its oversized Nuclep heavy components facility by gaining an order to assemble the lower part of the pressure vessel for Atucha-2, a transaction facilitated by Kraftwerk Union's status as foreign contractor for both projects. The Indian government is assisting in hydropower projects in Nepal, has built two 120 MW thermal power units in Libya, and has launched over 120 joint ventures abroad, providing technical consultants and equipment, as well as grants and credits.[20] In competing for exports to other developing countries, the relatively advanced developing countries enjoy historical advantages over the industrialized nations. They are free from colonial taint, and may be presumed to have greater empathy for the development of a particular country, having recently overcome similar problems.

MAKING CHOICES

Choice is complicated by the differences between projects which make head-on comparisons difficult. For example, a coal-fired plant may be built more quickly, with less initial capital outlay, than a hydroelectric station, but it will also have higher fuel costs and may be more hampered by environmental regulations. Somehow different lead times, capacity factors, capital, fuel, and operating costs must be conflated into an index which permits each alternative to be compared directly to others. This index should also include those factors bearing upon national objectives. To do this, economists first measure both the investment required for each item over the lifetime of the project, and the project's output over its lifetime (including its scrap value), compensating for inflation. The difference between investment and output equals the undiscounted value of the project.

To make projects with different lead times comparable, the value of each is discounted over the period from initial outlay to plant retirement. This calculation yields the present discounted value (PDV) of a project. There are many methods of cost–benefit analysis, but all aim to bring grossly incomparable projects into a comprehensible index for comparison. Cost–benefit analyses aid rough judgments, but are too adulterated by unwarranted assumptions to eliminate the need for other decision criteria, even if politicians would let it. The value of these calculations hinges upon the accuracy of a number of forecasts: interest rates, cash outflow for the project, cash return from electricity sales, and inflation. For nuclear projects, these forecasts are complicated by subsidiary uncertainties. One relates to construction costs. Historically, even in the developed countries, nuclear power plants have often ended up costing two to three times more than originally projected.[21] This has also been true for the developing countries, as noted in the case studies of Argentina and Iran, and as shown for India in Table 3.6.

Cash returns depend upon how much power is generated and sold, which in turn depends upon the capacity at which the reactor is running. Capacity factor forecasts have proved as unreliable as cost projections. Of operating Third World reactors, Atucha is the shining example, having enjoyed one of the highest capacity factors of all reactors worldwide, around 80 percent since 1974, the year it began delivering power.[22] TAPS and RAPS-1 and -2, by contrast, have performed poorly, with a cumulative load factor through 1977 varying between 30 and 50 percent, compared to a worldwide average of 60 percent.[23] KANUPP has weighed in with a miserable 15.6 percent from 1976 through 1980. Reduced load means reduced revenues, increased capital cost per kilowatt installed (capital cost divided by output), and a cost–benefit analysis distortedly favoring nuclear power projects.

The cash returns also depend on the prices charged for the electricity. Often governments pursue rural electrification programs from political rather than economic motives. The beneficiaries are seldom in a position to pay much for it. Rural electrification is often expensive, requiring installation of long connections to a remote hamlet with simple lifestyles and little if any industry. Electrical supply creates the demand there. In such cases, rural electrification customers cannot contribute enough revenues to cover program costs, and the PDV sinks below zero, only to be restored by the inclusion of the national objectives in the social discount rate.

Nuclear planners commonly use computer models to project the optimal schedule to build nuclear and other sorts of power plants. One feasibility study may employ numerous programs, to project capital and operating costs, future demand, cash flow, power outage effects, maintenance schedules, and the optimal order for building different types of power station. The virtue of these models is their ability to assimilate numerous data and generate alternative scenarios. Their weakness is that they can only be as accurate as their human estimated assumptions. A Boston-based consulting firm, for example, based its electrical load forecasts for Java on the whimsically optimistic assumptions that the state electrical enterprise, PLN, would replace all captive generators; provide electricity to all new industries, commercial establishments, and households in present service areas; establish service to all areas where there is demand but no PLN connection; and satisfy the growing demand in these new service areas. This study also indicated that electricity growth rates would exceed GDP growth rates by an enormous factor of 3 or 4 to 1: a 7.3 percent GDP increase, for example, was equated with a 25 percent electricity growth rate.[24] In fact, the electricity-to-national income growth ratio invariably is less than 2 to 1, and in Taiwan has been forced below 1 to 1.

Developing country forecasts often are developed in cooperation with the IAEA, which adopted the Wien Automatic System Planning Package (WASP) to help perform its Market Survey for Nuclear Power in Developing Countries. WASP, based on a US Tennessee Valley Authority program, simulates seasonal power station operation, evaluates operating costs, calculates present value of operating and capital costs, and determines total system costs to the planning horizons. Its input data consists of load forecasts, existing system description, alternative power plant descriptions, definition of acceptable level of system reliability, loading order of plants, discount and escalation rates, capital costs, salvage value of plants, etc.[25] Estimating these inputs entails a large margin of error, which is compounded when all are incorporated in one model. In the Indonesian study, which relied upon WASP, '[i]t was assumed that minimum costs rather than net benefits would be the measure of merit' in choosing between power expansion programs (p. 71). This emphasis upon cost minimization perhaps constitutes WASP's worst shortcoming: its neglect of resource availability. This may have been justified in the United States, where the model was developed and where necessary resources abound, but certainly represented a gross oversight for the developing countries. In Indonesia, its use led to the conclusion that between eight and eighteen reactors ought to be built between 1983 and 1997, in order to minimize electrical expansion costs. The IAEA study never addressed, let alone answered, the vital question of how the manpower or funding needed to build even one reactor would be found.

These criticisms aside, it should be recognized that overpessimism can be as dangerous as overoptimism in planning. Perhaps nuclear projects are 'too' difficult for present resources, but if it always shies away from the difficult, a government might never escape from the box of underdevelopment. If capital scarcities definitively refute proposals for capital-intensive projects, then growth of stocks will remain stunted, and shortages will become self-perpetuating. The same logic could be applied to other resources.

A theoretical framework, explaining how decision-makers take on intimidating tasks, was provided by Albert O. Hirschman.[26] According to his Principle of the Hiding Hand, since man necessarily underestimates his creativity, it is desirable that he underestimate to a roughly similar extent the difficulties of the task faced, so as to be tricked by these two offsetting underestimates into undertaking tasks that he can, but otherwise would not dare, tackle. Similarly, the failure to internalize external costs can stimulate enterprise. Hiding Hand techniques can teach decision-makers how to take risks. One method is to make a project look easier than it is or, when difficulties cannot be hidden, to exaggerate its prospective benefits. Particularly applicable to nuclear and overall energy planning is the 'pseudo-comprehensive' program technique, wherein decision-makers dismiss earlier attempts as piecemeal, and pretend to more insight than is actually available, by drawing up a comprehensive program. Hirschman argues that development projects should combine 'trait-taking' and 'trait-making'. In other words, if a project is planned, built, and operated on the basis of certain negative attributes of the *status quo*, assuming them to be inevitable and unchangeable, it may miss important opportunities for beneficially changing these attitudes – it may even confirm or strengthen them. Project design should *take* some traits as temporarily unchangeable but *make* others, to improve the status quo. If, for instance, a project makes educated workers, the next project may not need to take as much imported skilled labor.

The extent to which traits can be made and not taken depends upon an economy's ability to absorb new work patterns and new technologies. Energy technologies must overcome technological, social, and market inertia before they become commercialized on a wide scale.[27] A country's absorptive capacity is a function of its resources, social flexibility, and receptivity to modernization. (The Islamic fundamentalist movements show how leaders can sometimes overestimate receptivity.) To facilitate absorption the Argentine government created the Service of Technical Assistance to Industry, which serves first, to channel nuclear technology from the CNEA to industrial use, and second, to apprise the Commission of the needs, possibilities, and limitations of Argentine industry.[28] The fruits of this effort were apparent at Atucha-1, where Argentine industry contributed nearly 100 items worth DM30.2 million ($9.4 million at 1972 rates).[29] For Embalse, the level of Argentine participation rose to over 50 percent. The target for Atucha-2 was about 60 percent. Politics also affect absorbability, since any large-scale, long lead time, or expensive technology can only be introduced in an atmosphere of stability. Otherwise, the addition of political to economic and technical uncertainties will erase the margin of confidence needed to convince investors that their time, effort, and money will not be wasted by a struggling government or its unsympathetic successor.

THE INDEPENDENT, DEPENDENT, AND NONNUCLEAR COUNTRIES

To assess the importance of economic objectives to the nuclear policies of the independent, dependent, and nonnuclear countries, a brief examination of available resources and expressed policies follows.

The independent countries require the most educated labor. The Indian case is exemplary. In 1945, the Tata Institute of Fundamental Research was established in Bombay. Nine years later, around 130 scientists and technicians joined together at a new, multidisciplinary research and development center in nearby Trombay. This is now known as the Babha Atomic Research Center (BARC), and has become the foremost research institution in the country, with a staff of around 10,000, of whom nearly one-third are scientists. A majority of the scientists and engineers there graduated from the BARC training school, founded in 1957.[30] Argentine nuclear scientists and engineers broke the 1,000 mark by around 1970, and the high general educational level (94 percent adult literacy) in the country reflects the European birth or descent of over 90 percent of the population. In fact, the proportion (29 percent) of the population aged 20 to 24 enrolled in higher education in Argentina is higher than that of any other country outside the industrialized world.[31] The high Israeli educational standards are well known. Pakistan and South Africa both have highly qualified enclaves of nuclear engineers, due to their colonial heritages and the high priority of the programs there. Nevertheless, many more scientists are trained than retained, due to the 'brain drain'. Qualified workers leave the country by the thousand, attracted by the better working conditions and salaries offered in the industrialized nations. Labor importation can offset the brain drain, but is constrained both by financial resources and by reluctance to risk political unrest provoked by their unwelcomed presence.

The land resources of the independents vary. Argentina, India, and South Africa have large uranium deposits. Israel and Pakistan do not, but both are widely believed to have surreptitiously obtained the uranium they desired, and Israel also has its uranium-from-phosphate project.[32] With respect to energy alternatives, coal and hydropower are to provide the bulk of increased supplies by 2000 in India and Argentina, respectively. Despite governmental braggadocio, nuclear power expansion has taken a back seat in both countries, as Table 3.5 reveals. The larger shares given nuclear power in the other three independents reflect partly the lower overall consumption levels projected, and partly the fewer energy resources available in Israel, Pakistan, and South Africa. In sum, though the independent countries as a whole seem well able to procure whatever uranium they require, the rich alternative resources of India and Argentina have contributed to the incentives there to limit the nuclear role.

This situation is mirrored in capital. Nowhere has its shortage *prevented* an independent from pursuing its central objectives, but it has limited the scale of commitment in all cases. Israel has yet to invest in a nuclear power station. Pakistan's first power reactor cost only $105 million, and was financed by soft loans from Canada and Japan.[33] The South Africans' current account balance (after interest payments on external debt) shifted from a $1.21 billion deficit in 1970 to a $1.54 billion surplus in 1978, and even though debt service as a percentage of exports of goods and services increased during that time, it remained relatively modest at 11.7 percent.[34] When it came to negotiating financial arrangements for the Koeberg reactors with the French, the South Africans were able to take advantage of their strong credit rating, by obtaining two 900 MWe reactors for $1 billion in 1976, on good terms.[35]

In Argentina, financial resources have been the weak link. Although continued economic uncertainty generally discouraged investment, shortages of foreign credit (when it was welcome) were not the main problem, because Argentine beef and grain exports, along with oil self-sufficiency, ensured a generally healthy and resilient trade balance. In their eagerness, reactor vendors and their home governments offered excellent financing terms. Both Siemens and AECL were supported by their governments through export credit concessions. Both paid a high price for their easy terms and acceptance of stiff, penalty-backed performance guarantees, sustaining losses on their contracts on the order of tens and hundreds of millions of dollars, respectively.

The main problem, instead, has been domestic financing. Foreign loans, though depreciating, have to be serviced and repaid. While domestic credit costs climbed, national income has stagnated. To get a grip on three-digit inflation, the Videla government attempted to batten down public spending. Economics Minister Jose Alfredo Martinez de Hoz objected to the 1980–95 CNEA plan on the grounds that the country could not afford to build ten reactors in that period, and he succeeded in reducing that number to four. The energy sector at that time was the top governmental investment priority, but within it nuclear energy was given only a subordinate role. Larger sums went to the state oil company ($405 million), hydro projects ($483 million), and the State Water and Power Board ($265 million) than to the CNEA ($152 million).[36] Commission officials must have argued that it would be wise to exploit Argentina's plentiful uranium resources and that the level of nuclear power construction envisaged represented the minimum necessary to save the CNEA and nascent Argentine nuclear industry from atrophy. The President concurred with this view.

The major resource availability differences between the labor, land, and capital availability for the dependent countries were sketched in Chapter 3. Among the countries with major nuclear power programs, independent and dependent, the discrepancy in labor resources is often derivative, flowing from disparate nuclear policy objectives. None has a surfeit of qualified labor, but many are able to siphon off a sizable portion of the skilled pool if the priority of nuclear development so commands. Naturally, population size makes a difference – India has the second largest population of scientists in the world – but despite populations less than 3 and 6 percent as large, respectively, Taiwan and South Korea have nuclear work forces comparable to that of India. Others suffer more acutely from skilled labor shortages, especially when difficult goals are set. For instance, to fulfill the Shah's objective of 23,000 MWe, an estimated 15,000 skilled technicians were needed.[37] In 1976, approximately 150 AEOI employees had some nuclear physics background, and 90 percent of these were foreigners. Even with the hundreds of students in training at home and abroad, finding enough Iranians for the program promised to be a burdensome task. The needs of the program were so great that the AEOI could not afford the CNEA approach of training in basic fields, so the spillover of improved technical capabilities into other sectors would have been reduced.

More important are the differences in available natural resources, which often work at cross purposes. A country rich in alternative energy resources, such as Argentina, is encouraged to go the independent nuclear route by the government's confidence that it need not scramble for quick energy supply fixes, but rather can thoughtfully select the most attractive, long-run alternative. Nuclear power seems attractive in this light, especially to the oil-rich nations which not only have the luxury of leisurely policy choice, but also the petrodollars to bring any alternative, nuclear included, within reach. Iran and Mexico were the prime examples here. On the other hand, the very margin of safety which facilitates a nuclear decision depends upon those alternatives being exploited, and they may appear so economically competitive as to squeeze nuclear completely out of the short-term picture, as happened in Indonesia and, in 1979, Iran. A dearth of energy alternatives also cuts both ways. On the one hand, energy scarcity promotes nuclear power as a substitute energy source. On the other, painfully expensive oil bills strip governments of the large assets needed to pay for it. These cross-cutting factors help explain why the countries most committed to the dependent nuclear development route have included large energy exporters, (Iran and Mexico) as well as large energy importers (Brazil, Philippines, South Korea, Taiwan).

The capital side of the economics of nuclear power is the most difficult to fathom. A wealthy country, like Iraq or pre-revolutionary Iran, has a strong incentive to convert some financial into physical capital, especially when financial capital depreciates in real terms, as inflation outpaces returns on investment. For such a country, building a nuclear plant translates paper assets into goods, alleviates domestic pressure on domestic oil resources better saved for export and petro-chemical purposes, and prepares for the post-oil era. Even in such an extreme case, however, other considerations may contradict this logic. Despite the country's wealth, for instance, inadequate physical capital forced Iranian industry into a negligible role in the nuclear plan. AEOI Director of Power Plants, Ahmad Sotoodehnia, concluded that 'the main problem right now we are facing in the implementation of our power plant programme is . . . the non-availability of the technical manpower, the non-availability of the support industry inside Iran to contribute to this industry.'[38] This problem, when coupled with the Shah's premium on building nuclear power stations with utmost dispatch, created a vicious circle: because participation of Iranian industry would slow reactor construction, it was excluded. Because Iranian industry was excluded, it could not easily improve its ability to contribute to reactor construction.

In the other oil-exporting, capital-surplus countries of the Persian Gulf, insufficient demand for electricity discourages nuclear invest-ment. Even in Iran, where the nuclear option was embraced, the planned use of the electrical output of Bushehr remained unclear and financial headaches abounded. Costs of all four reactors under con-struction escalated enormously, for two reasons. First, although Iranian income arrived mainly in US dollars, the German and French reactors were paid for in the home currency of the contractors. Conse-quently, the deterioration of the dollar, especially against the deutsch-mark, hurt the Iranians. Second, inflation was serious in France, and was passed on to Iran. Critics of the program claim that the price of the Bushehr reactors increased from DM7,800 million to DM13,200 million, representing a per kilowatt cost over $3,000, and alleged that the per kilowatt costs for the French reactors were higher still.[39] This was triple the cost reported by AEOI when the contracts were signed. While there remains great uncertainty over the precise figures, that there was extreme cost escalation cannot be doubted. Meanwhile, the product of Iran's reduced oil revenues during this period was a large government deficit, which reached $15 billion in 1978, or a third of the total budget.[40] When it became necessary to make major budget cuts, in October 1978, the Shah froze new orders for nuclear reactors.

Nevertheless, extensive capital assets were the factor most conducive to a large Iranian nuclear power program. Short-term cash flow problems did not alter the long-term earning potential of Iran's petroleum, so financing large capital projects should have continued to have been feasible, if they had been well chosen. The important question is whether nuclear power could have survived the scrutiny of a careful selection process. Had Etemad's argument, that conventional power generation costs would inevitably climb relative to nuclear power, been true, then investment in the latter would have been reasonable. But the costs at Bushehr and Ahwaz themselves were climbing rapidly, and nuclear power costs had not yet stabilized even in the advanced nations. Other governments faced the large nuclear investment burden with far less capital, and indeed with large external debts requiring ever-increasing debt service payments, as shown in Table 2.3. Meanwhile, the costs of nuclear power stations continued to escalate, sometimes astonishingly so, as when an eighteen-month construction delay inexplicably jolted the price tag on the Philippine reactor from $1.1 billion to $1.9 billion.[41] Climbing costs and external debts need not halt projects underway, but they have dampened nuclear plans, and dissuaded governments who have yet to take the plunge to hold off at least until the situation stabilizes.

This final point brings the discussion around to the nonnuclear countries. Of these, some have more skilled labor (Venezuela), oil or uranium (Nigeria and Niger), or capital (Kuwait) than many of the nuclear countries, but none compare favorably in all these categories. On the contrary, most African and Asian nations compare dismally in labor and capital categories, while enjoying much better ratios of natural resources to current and projected demand. The contrasts between nonnuclear Latin America and the nuclear nations are less stark. Most nonnuclear countries suffer from an energy crisis of receding forests and increasing fuel prices, but large-scale, electrical networks sustained by nuclear plants cannot quickly alleviate the problem. In the short run, diversion of resources from present consumption or from short-term, partial solutions would only exacerbate existing hardship. In such circumstances, long-term planning becomes academic.

Those governments which enjoy certain economic strengths have to make a tougher decision: can a surplus of one factor (such as oil) compensate for shortages of capital or manpower? Most have decided that the factors were not infinitely interchangeable, and the Iranian case confirms the wisdom of that view. The Shah thought that his capital surplus could make up for other inadequacies, that he could simply import equipment, foreign technicians, and (if necessary) uranium, to ensure the success of his nuclear program. In his optimism, however, he disregarded the saturation point of social and economic absorption. Due to inadequate physical and human resources, as well as the Shah's impatience, the program could not be designed to prod economic growth, through steadily increasing electricity supply and dispersing new knowledge through many sectors. Instead, it was designed to catapult the Iranian economy into modernity, though precisely how such rapid construction of turnkey stations would have translated into general development remains unclear. What is clear is the fact that this was a strategy concerned with the extremely long term.

The Indonesian case was more typical of the nonnuclear country approach: a less ambitious, shorter-term perspective. Because Suharto's security objectives were satisfied by a small research program, the economic case had to carry the whole burden (as in Iran) of justifying a nuclear power program. In both countries, the lack of physical or human resources prevented successful advocacy of nuclear power as a good way to use indigenous resources or skilled manpower, which were in great demand in all sectors. Consequently, both economic cases for nuclear power were based on the premise that large increases in electricity production would catalyze economic growth. Since economic benefits increase with the scale of the units and the grids they supplied, one can better understand the anomalies of officials in countries which were much less able than Argentina or India to support a nuclear power program, advocating the installation of many more reactors. Indonesia, however, differed from Iran both in possessing less wealth per capita, and in not being ruled by a monarch devoted to nuclear power.

Financially, although Indonesia enjoyed a favorable balance of payments and strong foreign reserves, due to its oil exports, investment for a nuclear power station would have been difficult, even with good exporter credits. Throughout the 1960s, the Indonesian economy was far too weak to support such an investment. As the economy revived in the early 1970s, the prospect became more feasible. The willingness of the government to undertake a several hundred million dollar or more investment, most of which went to foreigners, diminished markedly in the world recession-induced reductions in export revenues and Pertamina's inability to meet international debt obligations. A country with so many energy alternatives could safely choose a more restrained energy investment scheme.

In all developing countries, decisions to go or not to go nuclear have been explained primarily in economic terms. The arguments on either side closely resemble those suggested in this chapter and in Part One, and need not be repeated country-by-country. The sophistication of most nuclear planning studies varies according to the resources of the government and the importance of convincing decision-makers that increasing power supplies is necessary to growth and that installing nuclear power is the best way to do it. In Argentina, the CNEA itself conducted the multivolume Atucha-1 feasibility study, which helped defeat the opponents to the project. In Indonesia, after a twenty-volume study commissioned by PLN to British consultant engineers gave short shrift to the nuclear option, at least until the mid-1990s, nuclear advocates eventually found a government (Italy's) willing to put up the money for a counterstudy, which reported favorably on the feasibility of nuclear power for Indonesia. The battle of the studies continues, the latest entrant being the powerful, US-based Bechtel National, Inc. weighing in on the side of the Ministry of State for Research and Technology.[42]

By contrast, neither Iran nor many nonnuclear countries have adopted the decision-by-study technique. In Iran, once the Shah had made up his mind, no study designed to assess the feasibility of nuclear power for Iran could matter, though it could delay the project, and therefore the formality was forgone. Nor did feasibility studies underpin the nuclear programs of India, Israel, Pakistan, Libya, or Iraq. In these nations, nuclear policy expressed powerful leadership objectives, not seriously opposed by important interest groups. In nonnuclear countries without nuclear power lobbies, nuclear planning studies were equally useless. These examples, drawn from the independent, dependent, and nonnuclear categories, converge to show that studies are agents of advocacy more than of impartial selection, no matter how heavily the underlying assumptions are disguised in statistical charts and multiple regressions. Where there is little conflict, studies play little roles, not because absence of conflict indicates the patent superiority of one choice, but because without serious opposition, the powers that be have little reason to justify themselves.

CONCLUSION

Where markets are defective, economic decisions become politicized. Where projects are technically difficult, political decisions become economized. Nuclear power policy therefore flows from a dynamic interaction between political and economic considerations. In the South, economic objectives rest on a broad consensus; everyone wants national income to grow. Political objectives are far more sectarian. The best way to promote nuclear power, then, is to aim for the consensus and demonstrate that it helps economic development. That is why officials whose political beliefs shape their views about nuclear power still defend their views on economic grounds. Environmental arguments matter, but sometimes are discounted, while issues such as the possible intrusion of the nuclear bureaucracy upon civil liberties appear hopelessly esoteric in countries where popular rights are often rudimentary.

Battles are fought primarily on economic turf for another more obvious reason: nuclear power is expensive. The political pros and cons to nuclear power are dwarfed in importance by the $1 billion price tag of a single reactor. Since many benefits can be gained by rhetoric or research-scale programs, political justifications become less cogent as program costs climb. It is not surprising that all countries with nuclear *power* (as opposed to research or weapons) programs rely fundamentally upon economic rationale to present their case publicly. By contrast, the Israelis, who as yet have no power program, have not pinned the case for their nuclear program on economic grounds. In other countries where nuclear power exists and where ulterior motives are widely suspected, such as India and Pakistan, officials religiously profess their benign intentions. Although some of this insistence may be disingenuous, economic rationales are not merely cosmetic. Economic changes can drastically alter the array of supporters and detractors. When the Indonesian economy suffered the double shock of deteriorating export revenues and the Pertamina collapse, and no marketable uranium deposits were announced after a decade of French assistance, the case for nuclear power sagged badly. Cost escalation corroded support for the nuclear projects in the Philippines and Iran. Conversely, the 1979 Iranian chapter of the oil crisis reinforced the perceived need to develop alternatives, including nuclear. On the whole, economic objectives are important determinants of a government's nuclear decision, but the direction of their influence hinges upon assumptions that mix politics and economics. The pronuclear lobby wants a powerful, industrializing economy and believes that energy supplies must increase in step with GNP. The antinuclear lobby wants equality, environmental conservation, and balanced development, and believes that appropriate or intermediate technologies should be emphasized. This is an oversimplification, but a rough distinction between the supply managers and demand stimulators can be useful. In what can become a frustratingly arcane debate, PDV cost–benefit analysis can sift out the egregiously nonsensical, but can only help, not replace, political decision.

NOTES

1   Aristotle, *The Politics*, trans. T. A. Sinclair (Harmondsworth, Middlesex: Penguin, 1962), pp. 27–8.

2   See I. M. D. Little and J. A. Mirrlees, *Project Appraisal and Planning for Developing Countries* (London: Heinemann, 1974), Chapter 4.

3   W. Walker and M. Lönnroth, in *The Viability of the Civil Nuclear Industry* (London: The Royal Institute for International Affairs, 1979), p. 31, noted that nuclear power, once installed, becomes profitable in inflationary periods due to the effects of historic cost accounting in reducing the real level of annual capital charges.

4   Atomic Industrial Forum, *Engineering Evaluation of Nuclear Power Reactor Decommissioning Alternatives* (Washington: Atomic Industrial Forum, 1976).

5   United Nations Industrial Development Organization, *Guidelines for Project Evaluation* (New York: United Nations, 1972), pp. 117–38.

6   For a detailed discussion of energy, economic growth, and development strategy, see J. Dunkerley *et al.*, *Energy Strategies for Developing Nations* (Baltimore: Johns Hopkins University Press, 1981), Chapter 4.

7   World Bank, *World Development 1980*, pp. 142–9.

8   United Nations, *Statistical Yearbook 1968* (New York: United Nations 1969), pp. 362–6, and *Statistical Yearbook 1976* (New York: United Nations, 1977), pp. 386–90. The developing world totals were calculated by combining Africa, Asia, and South America, then subtracting Japan.

9   R. Alemann, 'Economic development of Argentina', in Committee for Economic Development, *Economic Development Issues: Latin America* (New York: Praeger, 1967), p. 41.

10 B. J. Csik, 'Posibilidades de centrales nucleares de pequeña o mediana potencia en la Argentina', in IAEA, *Small and Medium Power Reactors. Conference Proceedings* (Vienna: IAEA, 1961), pp. 432–3; and *Frankfurter Allgemeine Zeitung*, January 22, 1974.

11 IAEA, *Nuclear Power Planning Study for Indonesia* (Vienna: IAEA, 1976), p. 61.

12 Richard J. Barber Associates, Inc., *LDC Nuclear Power Prospects, 1975–1990: Commercial, Economic and Security Implications*, ERDA-52 (Washington: US Energy Research and Development Administration, 1975), p. I.2.

13 IAEA, *Nuclear Power Indonesia*, op. cit., p. 61; US, Department of Commerce, *Iran: A Survey of US Business Opportunities* (Washington: US Government Printing Office), p. 108; J. L. Alegria *et al.*, 'La contribución de la energía nuclear a la solución del problema energético argentino', in United Nations, *Peaceful Uses of Atomic Energy: Proceedings of the Third International Conference*, vol. 1 (New York: United Nations, 1965), p. 106.

14 J. A. Sabato, 'Atomic energy in Argentina: a case history', *World Development*, vol. 1 (August 1973), pp. 23–7.

15 Taiwan Power Company, *Taipower and Its Development* (January 1980), p. 5. Amovy Lovins argues that T&D account for between one-third and two-thirds of the cost of producing and delivering electricity to large industrial and residential–

commercial consumers, respectively, in *Soft Energy Paths* (Cambridge, Mass.: Ballinger, 1977), pp. 87–8.

16   Of the remaining one-fifth, civil construction comprises 15 percent; erection and commissioning services, 17 percent. H.-K. Reinhold, 'Transfer of technology', in *Problems Associated with the Export of Nuclear Power Plants* (Vienna: IAEA, 1978), pp. 157–60.

17   Ibid., p. 165.

18   *Latin American Weekly Report*, March 27, 1981, p. 9.

19   *The Energy Daily*, September 24, 1980, pp. 3–4; and *Latin American Weekly Report*, March 27, 1981, p. 3.

20   *India News*, May 14, 1981, pp. 8–10.

21   This often resulted from competitive bidding down of prices to win the contract in the first place. See I. C. Bupp and J.-C. Derian, *Light Water, or How the Nuclear Dream Dissolved* (New York: Basic Books, 1978).

22   Definitions of capacity factor, load factor, operating load factor, and availability are sometimes chosen or manipulated by reactor operators to show their performance in the best possible light. See the dispute over KANUPP's performance in *New Scientist*, February 12, 1981, p. 403, and March 19, 1981, p. 765, where a proponent's estimate of 74 percent availability was criticized by a critic who stressed that load factor was only 15.5 percent in that period.

23   R. Tomar, 'The Indian nuclear power program: myths and mirages', *Asian Survey*, vol. 20, no. 5 (May 1980), pp. 524–5. Similar data were reported by Dr M. R. Srinivasan, Department of Atomic Energy, in *India News*, October 9, 1980, pp. 4–5, giving TAPS a 54 percent rating from 1970 to 1980.

24   Chas. T. Main International, Inc., *Planning Study for the West Java and Central Java Interconnected Systems* (Boston, Mass.: October 1974), cited in IAEA, *Indonesia*, op. cit., pp. 63–7.

25   R. T. Jenkins and D. S. Jay, *Wien Automatic System Planning Package (WASP): An Electrical Utility Optimal Generation Expansion Planning Computer Code*, US AEC Rep. ORNL-4945 (July 1974), referred to in IAEA, *Nuclear Power in Indonesia*, op. cit., pp. 72–3, 141–67.

26   A. O. Hirschman, *Development Projects Observed* (Washington: The Brookings Institution, 1967), *passim*.

27   For discussion of the constraints to various energy supply options, see J. Anderer, A. McDonald, and N. Nakicenovic, *Energy in a Finite World: Paths to a Sustainable Future* (Cambridge, Mass.: Ballinger, 1981), Chapter 7.

28   Technical Assistance Service examined in J. A. Sabato *et al.*, *Energía atómica e industria nacional* (Washington: Organización de los Estados Americanos, 1978); A. Aráoz and C. Martínez Vidal, *Ciencia e industria: un caso argentino* (Washington: Organización de los Estados Americanos, 1974).

29   J. N. Baez *et al.*, *Participación de la industria argentina en la central nuclear en Atucha y futuras* (Buenos Aires: CNEA, 1973).

30   H. N. Sethna, 'India's atomic energy programme – past and future', *IAEA Bulletin*, vol. 21, no. 5 (October 1979), pp. 2–3.

31   World Bank, *World Development 1980*, pp. 154–5.

32   Libyan leader Qadhafi apparently purchased yellow cake from Niger and shipped it to Pakistan until 1979, *Washington Post*, December 23, 1980, p. 15. The Israelis allegedly hijacked the Scheersberg, a ship which left Rotterdam with 200 tons of uranium ore in November 1968, bound for Genoa under a Liberian flag and with a British crew. Later the ship reappeared with a different name, but the uranium was gone.

33   P. B. Sinha and R. R. Subramanian, *Nuclear Pakistan: Atomic Threat to South Asia* (New Delhi: Vision Books, 1980), pp. 33–4.

34   World Bank, *World Development 1980*, p. 135.

35 International Institute for Strategic Studies, *Strategic Survey 1976* (London: IISS, 1977), p. 115.

36 *Business Latin America*, February 14, 1979, p. 51.

37 E. W. Lefever, *Nuclear Arms in the Third World* (Washington: The Brookings Institution, 1979), pp. 54–5.

38 A. Sotoodehnia, quoted in AEOI, *Transfer of Technology*, vol. 4, p. 565. See also F. Halliday, *Iran: Dictatorship and Development* (Harmondsworth, Middlesex: Penguin, 1979), p. 147.

39 B. Mossavar-Rahmani, 'Iran's nuclear power programme revisited', *Energy Policy*, vol. 8, no. 3 (September 1980), p. 199.

40 Ibid., p. 190.

41 Philippine officials would not provide detailed breakdowns, but estimated that added safety features accounted for only $70 to $80 million of the increase.

42 BPP Teknologi, in association with Bechtel National, Inc. *Alternative Strategies for Energy Supply in Indonesia, 1979–2003* (Jakarta: December 1980).

Domestic Politics

Policies emerge through conflict and compromise. Sometimes they accurately reflect the wishes of a president or prime minister; other times, bureaucrats frustrate leadership intentions. Where leadership cannot impose its will on bureaucracy, or bureaucratic jockeying for status and influence distorts leadership desires, policies may stray far from original objectives. In such a situation, the security and economic objectives discussed in the last two chapters could influence nuclear policy less than first appears. Of course, *any* nuclear policy is a product of domestic politics. The purpose of this chapter is to determine the extent to which domestic political processes circumvent leadership desires, if at all.

A popular mode of thought, sometimes known as the bureaucratic decision-making model, rejects the notion that policies represent the considered outcome of a rational process, where policies emerge as a considered choice of the optimal solution among various alternatives.[1] Rather, the argument continues, policies often represent the only workable compromise among turf-jealous bureaucrats and between political parties more interested in defeating opponents than in seeking wise policies. Decisions percolated through this sometimes haphazard process may end up reflecting no-one's first choice, or everyone's second, but in any case something different from leaders' initial security, economic, or any other objectives.

This chapter argues that this model of decision-making does not generally apply to developing country nuclear policy-making. To be sure, bureaucratic arrangements heavily influence nuclear policy, but they do not relentlessly bend it out of all semblance to national objectives. This does not deny the validity of the bureaucratic decision-making model elsewhere, but only for decisions relating to major projects in countries with comparatively small and simplified policy-making apparatus. A corollary flows from this argument, detailing how major projects confront leaders and bureaucrats alike with a *dilemma of scale* which seriously complicates successful decision-making.

Two reservations apply to this discussion. First, graft is not considered. Charges of kickbacks and commission fees have arisen in regard to most nuclear power station contracts signed in developing countries, and sometimes have been admitted or proved. Their omission, however, should not damage the conclusions of this study. Tracing these transactions requires detective work and speculation, which could distract from the specific study of nuclear policy-making. Since graft is common to many large energy investment projects, there is no reason to assume that it is more important to the approval of a nuclear power program than to approval of coal, oil, hydro, natural gas, or other projects. Second, this chapter is concerned primarily with principal policy-makers. Legislatures play a nominal role (the AEOI was formally established by the Majlis, Congressional inaction killed a 1975 CNEA reorganization plan); court involvement is possible (in suits arising from alleged radiation exposure damages); and political parties occasionally speak out (the Peronists attacked the heavy water plant contract award to the Swiss firm, Sulzer Brothers); but these seldom determine major policies. Public support is encouraged and dissent avoided, but popular involvement in decisions is even more sporadic and peripheral. For example, there is no evidence that the public debates surrounding the proposed Embalse and Atucha-2 reactors in any way altered the contract award decisions. In the Philippines, popular opposition was able to vent its protests through a review panel established in June 1979 by President Marcos in response to concerns generated by the Three Mile Island accident. The president even suspended construction on the Westinghouse reactor under construction on Luzon. Nevertheless, Marcos decided three months later to resume the project, leaving the nuclear opposition helpless. In India, the government seems deaf to complaints that the Tarapur station exposes its workers to excessive radiation doses, and has not shut down TAPS for a thorough decontamination because it is reluctant to reduce electrical output.[2]

POLITICAL LEADERSHIP

Can one man move an entire nuclear program and, if so, how far? The history of atomic energy policies in developing countries suggests that one man can make an enormous difference. The premier example was India's Dr Homi Babha, a powerful promoter who enjoyed the confidence of Prime Minister Nehru and the esteem of the international community. Under Babha's stewardship, beginning in 1945, a country steeped in poverty mounted a large nuclear research effort, with one research center alone employing 1,800 professional scientists and engineers, 3,000 technicians, and 3,200 support staff by the time of his death in a 1966 air crash.[3] By the early 1960s, the Indian Atomic

Energy Commission had built three of its own research reactors, as well as fuel fabrication and reprocessing plants. Admiral Quihillalt transformed the Argentine program from a modest research effort into a viable power program. The stagnation of the CNEA under Admirals Helio-Lopez (1958–60) and Iraolagoitia (1973–76) demonstrated the palpable impact of the arrival and departure of one man. In four years, Etemad built an organization of 4,000 from the ground up and initiated the construction of four large power reactors.

No matter how capable or dynamic, atomic energy promoters rely enormously upon political leaders. Political systems in the developing world feature strong executives, whose support is indispensable for the development of technology as expensive and complex as nuclear fission. Governmental budget battles are inevitable. A head of state hears numerous lobbies. If he will not intervene on behalf of a nascent nuclear program, then he commits it to suffer at the hands of competitors. Thus, in the late 1970s, BATAN's ambitions were trampled by budget planners and coal developers. In Mexico, the administrative wrangling which has delayed the Laguna Verde project by several years might have been reduced if not eliminated by firm presidential intervention.

The Shah of Iran's unstinting support made his atomic program virtually impregnable in Cabinet disputes. The Iranian example illustrates the limits of effective leadership support of atomic energy development. The Shah could override opposition, but he could not prevent it. When important governmental factions begin to dissent from a program and sponsor studies to back their case, the leadership must take note. The antinuclear lobby seeks press coverage, and eventually the issue becomes a public issue, forcing the leadership to spend political capital if it wishes to maintain the program. By this time, the government's position may have become irretrievably awkward. Abandoning the project under duress demonstrates weakness, and tacitly admits that the critics were right, with the unconcealable implication that money and manpower have been wasted on an ill-conceived prestige project. Continuing the project risks increased criticism of the nuclear project, which may at some unpredictable moment spread beyond the nuclear sphere into a condemnation of general governmental incompetence or profligacy. Meanwhile, mounting expenditures on the nuclear project steadily refuel the critique. No graceful escape remains.

Leaders try to avoid this impasse. Tension inevitably arises between the mutual dependence of leaders and atomic energy commissions in promoting nuclear power, and their divergent interests in maintaining political power. Political leaders and atomic commissioners alike desire first and foremost to maintain their influence, but before a

nuclear program is underway the commissioners have little leverage
with which to persuade the leaders to take the plunge. The fortunes of
an atomic energy commission improve with a growing nuclear commitment. Political leaders look to a broader base of support. Nuclear
power is a difficult and chancy endeavor, and the natural political
aversion to risk places an executive at loggerheads with atomic energy
lobbyists. The politician's problem is complicated by the disparate
desires of the various groups he relies on to stay in office. A general
admires the sophistication and prestige of nuclear power, an
economic planner suspects its economics, and the populace marvels at
it, but may also fear its possible effects. The commission's best
recourse is to argue that nuclear power is essential, first, to national
prosperity and, second, to avoid falling behind other nations. It must
also seek as many allies as possible in trying to coax the initial government commitment. Once that Rubicon has been crossed, the political
leader's fate is to some extent linked to that of the nuclear program,
and the influence of the atomic energy commissioner rises accordingly.
Thus were the succession of Argentine presidents reluctant to remove
Quihillalt (after Frondizi's bad experience) or otherwise to tamper
with the highly visible and apparently successful atomic program.

President Marcos showed how a president may preemptively imperil
a nuclear program in order to protect himself politically. Excavations
began in February for the first nuclear unit, PNPP-1, at Morong on
the Bataan Peninsula, around 70 kilometers west of Manila. By late
1978, the project already was subject to severe criticism. Under
investigation by US Congressman Clarence Long, Westinghouse
admitted having paid $35 million to Herminio Disini 'for assistance in
obtaining the contract and implementation services'.[4] Then, it so
happened that a Filipino ex-patent lawyer, Augusto Almeda-Lopez,
was living in Harrisburg when the TMI-2 incident occurred. He wrote
to Senator Lorenzo Tañada that while the Americans had coped with
the crisis through cooperation among governmental, academic, and
industrial resources, the Filipinos could not count on such prompt,
capable, and extensive support if a similar emergency occurred at
PNPP. Taking the initiative, on May 15, 1979, two weeks after
TMI-2, Marcos issued Executive Order No. 539, which established a
special commission to investigate the dangers that might arise from
the existence or operation of PNPP-1. Marcos asked Assemblyman
Ricardo Puno to co-chair this commission, instructing him that 'we
cannot afford the *merest doubt* where the future of an entire nation is
concerned'. Two months of daily hearings followed, during which 623
concerned citizens of Bataan appealed to Marcos to cancel the project,
the US Department of State froze nuclear exports to the Philippines,
and a previously unreported seismic fault was discovered 400 to 500

meters from the site, which had already been criticized as too seismically and volcanically active for safety.[5] Marcos prepared a scapegoat by threatening to terminate the contract if Westinghouse failed to secure a US Nuclear Regulatory Commission (NRC) export license. The Puno Commission found much to criticize in Westinghouse's performance, but not in the basic choices made by the Philippine government concerning the safety of a nuclear reactor sited in the 'Pacific Ring of Fire' and the wisdom of spending over $1 billion for a 600 MWe nuclear unit. In May 1980, the NRC finally approved the export license needed to ship parts to PNPP-1; in September Marcos permitted work to resume on the project.[6] Whether he was truly shaken by TMI-2 or cannily exploiting the opportunity to defuse nascent discontent with the project, Marcos clearly demonstrated the heavy hand presidential authority lays on nuclear power.

Political infighting increases with the stakes; a large, long lead time project is especially vulnerable, particularly in its early stages, before it becomes politically, financially, or literally set in concrete. It absorbs revenues which are coveted by other project promoters and parted with reluctantly by taxpayers. A research reactor can be built in a year, but a power reactor requires at least six from start to finish. Consequently, the stability of the political leadership affects its ability to sustain a successful power program. The Indian atomic program was run from its inception in 1945 by Babha until his death twenty-one years later. Since 1971 that role has been filled by Honi Sethna. Quihillalt's tenure spanned eighteen years and eight governments. On the other hand, despite nearly twenty years in office, Indonesian nuclear chief Siwabessy achieved far less, even accounting for scale. The lesson is that while stability of political leadership is not necessary (strength is), stability of atomic energy commission is not sufficient, for a successful atomic energy program.

An atomic energy commissioner must be a persuasive advocate in order for his voice to carry over the din of competition. When his proposals meet approval, he had to be able to execute them effectively in order to accumulate a record which can justify future expenditures. Pragmatism fares better than ideology: Quihillalt had minimal concern for unorthodox political views so long as they were not openly expressed. When Peronists reversed this policy in 1973, the results were unfortunate. (Perhaps they simply tried to do too much too fast, in response to the understandable frustration at having been out of power for so many years.) BATAN is vested with certain responsibilities for nuclear power which Chairman Baiquni zealously guards. This has lost him the sympathy of potential allies, such as geologists interested in uranium exploration and PLN officials desirous of involvement in nuclear power. Greater flexibility could facilitate the alliances necessary to bring nuclear power to Indonesia.

Individuals can be effective in these countries because of the nature of the policy processes there. Often only a few people need to be convinced of a policy's value for it to be approved, since key decision-makers commonly belong to a ruling elite, sharing similar cultural values, training, and outlook, which help smooth areas of friction. In Iran, for example, the natural antipathy which often develops between energy ministers and atomic energy commissioners was moderated by the good personal relationships between the leaders of each agency. At the level of implementation, leaders can rely on more members of their elite, many of whom have advanced degrees from the advanced nations. The effectiveness of these executive teams depends upon the depth of talent upon which they, in turn, can rely.

BUREAUCRATIC ORGANIZATION

Bureaucratic organization initially reflects the political values of its architects. Exceptional success of a new bureau, while others flounder, may reflect neither unfair treatment of the less successful nor distortion of the overall blend of policies. Instead, the political leadership may want a program of exceptional importance to be executed by an exceptionally strong bureau, as when nuclear power programs are administered by powerful, independent agencies. Conversely, changes in leadership or values can strip nuclear power of privileged status. Then atomic energy commissions have to battle much harder for success. At issue here is whether an atomic energy commission in a developing country can become strong enough to carry on effectively on its own, without strong political backing. This happened in France, where the Commissariat à l'Energie Atomique's work continued uninterrupted despite the succession of Fourth Republic governments throughout the 1950s. Even without a clear leadership decision to develop nuclear weapons, the CEA moved steadily toward nuclear weapon capability from 1952 on, without any express governmental sanction for a military program. By 1958, when the political argument for a French nuclear force became strengthened by the East–West balance of terror and deteriorating relations with NATO, the CEA was able to obtain Premier Gaillard's approval for a French nuclear weapon test in 1960.[7] French precedent, however, cannot apply in the South. Funds and skilled personnel are too scant in the developing countries for either to be secretly siphoned off to a program without explicit leadership backing. The following section explores the nature of developing country atomic energy commissions and their bureaucratic competition.

ATOMIC ENERGY COMMISSIONS

Two salient features mark most atomic energy commissions: specialization and autonomy. A single agency rarely shares atomic energy development with responsibilities for fossil fuels and hydropower. Occasionally, responsibilities for 'exotic' technologies, such as solar, were vested in an atomic energy commission, usually to the detriment of the addition. In Iran, AEOI was responsible for solar development for a time. Specialization enables a commission to devote all its vigor to the pursuit of a narrow set of goals. Most atomic energy commissions are also autonomous. Autonomy, more than specialization, aids advocacy. Atomic development requires full commitment, in economies as strained as those of the developing world, for programs might easily succumb to continued, energetic scrutiny. Subordination introduces policy guidance and budget allocation from a minister responsible for many programs, including some with objectives counterproductive to atomic energy development. In its most powerful form, the commission is a direct agent of the head of government, to whom the commission chief has direct access. If the head of state backs the commission in resisting budget cuts, then a planning agency can do little. Whether their specialization and autonomy provide enough leverage for atomic energy commissions significantly to affect the level of atomic energy in a country depends upon the relative strength of the bureaucratic competition.

Autonomy and specialization, however, are no panacea. First, they can only help at the margins. If the commitment to nuclear power is half-hearted to begin with, then autonomy and specialization contribute only to isolation, not to influence. Such has been the fate of many developing country atomic energy programs, relegated to a university physics department, with poor prospects for expansion. In Indonesia, the nuclear program was specialized and officially autonomous under Sukarno, but it remained so small that these advantages could not take it far. Four years after the establishment of the Institute of Atomic Energy, there were still only 100 trained personnel.[8] Even then, a former BATAN scientist said that there were more people than could be usefully employed due to the lack of projects. By 1970, the budget had not yet surpassed $300,000.[9]

Second, autonomous agencies are often established in order to escape from typical developing country problems which may prove inescapable. These have been categorized by Hirschman as inefficiency, inadequate salaries, frequent changes in key personnel and policies, nepotism, corruption, and 'politics',[10] faults often too deeply entrenched to be swept away by administrative fiat. Ironically, to the extent that the agency succeeds, it creates new problems for itself. In Iran, the richly endowed AEOI was able to attract good personnel by offering exceptionally good salaries, but in so doing incurred resentment of its privileges from other agencies. Resentment can easily translate into foot-dragging or other disruptive behavior, which can become a serious problem for an atomic energy commission dependent on others for finance, industrial assistance, electrical connections, and so on. Since the bureaucratic isolation arising from autonomy is indiscriminate, it cuts both ways, alienating important allies as well as preventing outside interference.

Physical, human, and financial resources can reinforce or undermine organizational advantages. CNEA's strong human resources contributed to the success that enabled it to deal effectively with competitors, while the 23,000 tons of domestic uranium reserves shielded it from attacks to which BATAN was vulnerable. The economic difficulties of the country highlighted the CNEA's Achilles' heel: its multibillion dollar program costs. The Economics Ministry, not surprisingly, became the most formidable combatant when the CNEA sought approval in 1978 for its fifteen-year plan. In Iran, abundant revenues permitted the Shah to embark upon a major nuclear program. Paucity of skilled manpower led to large-scale foreign hiring, which probably weakened the AEOI's case in its bureaucratic fights to maintain exceptionally high pay scales, though the Shah's personal support assured that the organization prevailed.

Overall, autonomy helps more than it hurts an atomic energy commission. Resentful bureaucrats can sometimes be mollified by the autonomous offender, as when Etemad actually had budget surpluses to bestow on the needy Energy Ministry. Alternatively, the bureaucratic antagonist may be so explicitly instructed by the leadership that to object or fail to comply would be to court dangers greater than those posed by the atomic energy commissions themselves.

Not all AECs hold the same writ. Some, like the CNEA, are fully responsible for research and development, training, power plant construction, regulation, and operation. Others, like the Mexican CNEN and the Taiwanese AEC, leave the power plant side of operations to the electrical utilities. Still others, like the Philippine AEC, are essentially confined to research and regulatory functions. In Brazil, a national company, Nuclebras, separate both from the CNEN and the electrical utilities, is responsible for all nuclear plant and fuel cycle manufacture. In its early stages, a developing country nuclear program usually lacks enough personnel to divide regulatory from research and promotional functions. It seems wasteful to label able nuclear technicians as 'regulators' and thereby lose them for more productive work.

Even the most specialized and autonomous AEC imaginable could not be viewed in a vacuum. The salience and expense of nuclear power guarantee the active interest of other parties, to whom we now turn.

OTHER ACTORS

Nuclear power programs generate opposition on economic and health grounds. Finance or economics ministries and planning agencies, responsible for paring budgets, can create serious problems for an atomic energy commission. They are often antagonistic toward nuclear power, for several reasons. To buy a station, governments have to deepen foreign debt and dispense foreign exchange. Also, because nuclear power is so capital-intensive, it provides fewer employment opportunities for each dollar invested than do labor-intensive activities.[11] Many employment opportunities in the nuclear field require more education than all but a few in the country have had, necessitating the employment of foreigners. Nuclear power development entails a long gestation period, and its promised returns seem remote and uncertain. Meanwhile, reactor prices continue to rise and doubts multiply over the economic viability of nuclear power for even the developed countries.

Though alternative energy technologies have often lacked the institutional base enjoyed by nuclear power, they are now beginning to come into their own. In Israel, a country with no oil, by 1981 one out of every three homes had hot water piped from simple rooftop solar heating panels, for a total of 1.5 percent of national energy needs.[12] Solar research absorbed $5 million annually. The program took on special urgency due to the Arab boycott, the return of the Sinai oil fields to Egypt, and the 100 percent reliance on oil to generate electricity, in the absence of any coal or nuclear stations. Israeli oil expenditures grew from $775 million in 1978 to around $2.25 billion two years later. Domestic solar research explored the use of saline water to increase heat absorption, chemical treatment of glass panels, and the use of liquid mercury impelled by heat through a magnet to generate an electric field.

In the Philippines, geothermal power receives high priority. In 1979, it accounted for 7 percent (220 MWe) of total installed Luzon power systems, a share which is slated to increase to 15 percent by 1989. The national energy plan envisions that marsh gas, biogas, wind energy, hot springs, direct solar, producer gas and dendro-thermal (wood) technologies will increase their collective contributions from 279,000 barrels of oil equivalent in 1980 to nearly 15 million barrels in 1989, or from less than 1 percent to nearly 10 percent of nonpower energy demand. 'Alcogas', produced from sugar cane and cassava, is itself expected to displace 20 percent of gasoline consumption in Metro Manila and selected areas of Luzon within a decade.[13]

The leader in the 'gasohol' field is Brazil, since 1972 the world's leading sugar cane producer. Alcohol, produced from sugar cane, has been used to fuel automobiles since the 1920s, and has ranged up to 8 percent of gasoline content in recent years. In 1974, the Commerce Ministry began research into raising the ratio of alcohol to gasoline and into burning pure alcohol in automobiles.[14] This program enjoyed the personal support of President Ernesto Geisel (who earlier headed the national oil company, Petrobras). Political attractions of the gasohol program include its relief of air pollution, job creation (up to one million, mostly agricultural), stimulation of capital goods production, and improvement of national self-confidence from a uniquely Brazilian solution to its energy crisis. The country is also rich in shale oil resources. A 1,000 barrels a day pilot plant has been operating there since 1972, and a 51,000 barrels per day commercial plant, which will also yield 1.8 million cubic meters of gas daily, is planned.[15]

All of these projects survive and thrive only at the sufferance of the political leadership. They are seldom born with powerful natural constituencies. Common sources of governmental support for nonnuclear alternatives include the environment and tourism ministries. Taiwanese and Turkish tourist authorities opposed construction of reactors in scenic venues. Similar concerns instigated opposition to Brazilian plans to locate the fourth and fifth nuclear power units near São Paulo, Emil Salim, Indonesian Minister of Environment and one of Suharto's senior advisers, in the aftermath of the 1979 oil boom remained the key opponent to the nuclear program there.

Most auspicious for new energy technology development have been increases in oil prices and in the effectiveness of ecology-minded promoters in foreign governments and international aid institutions. In the late 1970s, nonconventional energy prospects waxed while those for nuclear energy waned. Many alternatives do not compete directly with nuclear power in centralized electricity generation, but in a sense present a far more fundamental threat to fission programs, by making the latter appear irrelevant. Here we revert to the hard versus soft path debate; solar and other technologies are the favored choices of those who believe that 'elegant' solutions to energy demand management yield better results than gross increases in energy, especially in electricity supplies. Soft path theorists seek to build monied constituencies for these alternative technologies whose financial interests will move them to destroy the nuclear lobby.

Building interest groups is unnecessary in the well established fossil fuel and hydropower sectors. Oil companies seldom battle an atomic energy commission: many wish to minimize the use of oil for electricity generation by substituting nuclear power. They only threaten an atomic energy commission when seeking to ensure their continued dominance in the post-oil era, by diversifying into other energy technologies. For instance, a major 1973 seminar in Indonesia concluded that Pertamina should rectify the scattered approach to energy policy by assuming overall, coordinating responsibility for energy research and development, including nuclear. This idea was scotched two years later by the collapse and near bankruptcy of Pertamina.

Hydropower, which currently accounts for 44 percent of electricity output in developing countries, could be greatly expanded in many, especially in the great Amazon and Parana basins of Latin America.[16] Not surprisingly, the governments of Argentina and Brazil have staked the greatest share of energy investment resources on hydropower. In Argentina, the government decided to exploit most available hydropower resources by around 1990, while nuclear power was phased in gradually. Installed hydropower increased from 129 MW in 1956 to 1,745 MW in 1976, while two large projects begun in the late 1970s were expected to add 4,320 MW more.[17] The 1978 national energy plan there schedules installed hydropower to rise from 2,910 MW in 1978 to 19,840 MW in 1995, increasing its share of total electrical capacity from 31 to 66 percent.[18] Nuclear power is to contribute less than 10 percent installed capacity by that time. Brazilian hydro efforts similarly dwarf the country's nuclear program, and Nuclebras President Paulo Nogueira Batista told an investigating congressional committee that the nuclear budget to 1995 will account for 'only 15 percent of the total program for transmitting and generating electricity'.[19] Even in India, hydro outstripped nuclear power generation a dozen times over in the mid-1970s (33 TWh versus 2.6 TWh in 1975), although planners hoped that nuclear capacity would grow much faster, to the point where it would exceed two-thirds the hydro output by 2000.[20]

Many hydro programs have enjoyed stronger bureaucratic support than the nuclear competition, by virtue of their seniority and relative technological ease. In Argentina, for example, at the time of the Atucha-1 feasibility study, strongest competition came from 'a powerful group, backed by the Secretary of State for Energy, completely opposed to any nuclear power station, due to preference for exclusive hydropower development'.[21] The strong hydro lobby in the Energy Ministry persisted, winning top priority hydro development in the 1978 plan, and reportedly halving early CNEA requests to build ten nuclear stations by 1995. The hydro lobby had important allies in the CNEA, where key personnel feared that to try to build too many reactors too quickly could overtax their capabilities and jeopardize the entire nuclear effort.

Energy ministers sometimes arbitrate, sometimes advocate. An atomic energy commissioner must tread carefully in their domain, for even with explicit leadership blessing, a rancorous minister can quietly undermine nuclear efforts. Take the example of the Atomic Energy Organization of Iran, which enjoyed unsurpassed leadership support. In the first place, the Ministry of Water and Power did not sympathetically view its creation. The Ministry had been assigned to perform the 1972 nuclear feasibility study, and had been unenthusiastic then due to its preoccupation with conventional power sources and a belief that nuclear power was unrealistic and unnecessary for Iran. Trained professionals in the electrical field were scarce and officials doubted that the Ministry could diversify its functions to include nuclear power without compromising its effectiveness.

The foot-dragging encountered in the 1972 feasibility study showed once and for all that only an independent atomic energy commission could bring nuclear power to Iran. The AEOI's creation implied that the Ministry of Water and Power was incapable in the field and presented its executives with a powerful competitor. The AEOI, however, depended upon the Ministry (changed to the Ministry of Energy in 1975) and particularly upon its subsidiary, Tavanir, which was responsible for construction, transmission, and distribution of all nonnuclear electrical capacity. Tavanir would have been responsible for operating the reactors built through the AEOI. But relations between the two agencies were acrimonious. For instance, Tavanir was not extending T&D lines quickly enough to suit the Organization. The three 400,000 volt lines that were to carry the electricity produced at Bushehr lagged far behind schedule, raising concerns that, when complete, the two reactors would have had no outlets. Admittedly, Tavanir's tardiness stemmed partly from the many demands imposed on it from all over the country, but resentment of the AEOI's special status also contributed.

This example reveals the importance of the utility. Many oil-fired power stations remain in operation and will not be retired for decades. A few are still being built. Since 1973, however, utilities understandably have preferred to acquire stations using other fuels. Sometimes uranium seems a good replacement, but the electrical utility–atomic energy commission relationship is uneasy. Utilities are in business to generate and deliver electricity. Atomic energy commissions are in business to develop nuclear power. Commissions lack utilities' experience in running electrical power systems. Utilities lack commissions' familiarity with nuclear technology. Both want to maximize control over their own activities, including nuclear power. Alliances are possible, for nuclear power offers utilities a way to increase electricity production and utilities offer atomic energy commissions the opportunity to graduate from research to nuclear power. But these alliances are fitful.

Disputes between AECs and utilities are not confined to the developing world. In the late 1960s, the French CEA battled vigorously against the national utility, Electricité de France (EdF). The CEA wanted the French program to stick with the gas-graphite reactor it had developed since the early 1950s, while the EdF preferred to switch to the Westinghouse-designed pressurized water reactor, primarily because the American product was deemed cheaper and more reliable. The CEA lost that battle, and in 1969 was even threatened with extinction, though it has since recovered and retains control over fuel cycle and research, including the breeder program. The critical transition phase from a research to a power program determines whether the AEC will retain control over construction and operation, be deprived of those roles but retain responsibility for the fuel cycle, or be reduced to regulatory and research roles. Debates proceed on various fronts. In Argentina, opposition from the utilities centered around two issues: fuel cycle and reactor operation. First, Argentine utilities, like EdF, were less concerned with independence in the fuel cycle than with reliable delivery of increasing amounts of energy at reasonable costs. They preferred the light water reactor which, though it entailed dependence on foreign supply of enriched uranium, was cheaper and had a more extensive operating record than the heavy water reactor. If an expensive heavy water station performed poorly, it would have damaged utility efforts to persuade industries to buy their services instead of building their own generators.

Second, the Atucha-1 station was to be part of the Greater Buenos Aires–Littoral electrical grid built and run by the state-owned utility, Electrical Services of Greater Buenos Aires (SEGBA). Utility officials felt that they were better qualified than the CNEA to run a power station, and that dividing responsibilities for the grid could create confusion and inefficiency. Their critics labeled SEGBA a wasteful 'fossil', unfit to take on extensive new responsibilities. A commission to determine who should operate Atucha-1 decided that only the CNEA possessed the expertise to operate and maintain the first power reactor and, with the concurrence of the Secretary of State for Energy, authorized the Commission to take that responsibility. As the Embalse station neared completion, the issue of who was to operate and maintain it remained unresolved.

Interestingly, the balance of power between AECs and utilities parallels the division between nuclear independents and dependents. Atomic energy commissions, departments, or boards are dominant in Argentina, India, Israel, Pakistan and South Africa. Utilities dominate the nuclear scene in Mexico, the Philippines, South Korea and Taiwan. Brazil does not fit the pattern perfectly; its AEC is relatively small, though it has important responsibilities for training, research and regulation. The Institute for Atomic Research is the major state agency involved in nuclear R&D. Fuel cycle and reactor construction activities are carried out by majority government-owned nuclear and electrical companies, Nuclebras and Eletrobras, and by the electrical utilities. Nuclebras is charged with directing and implementing the overall nuclear energy program, coordinating the work of subsidiaries, and negotiating directly with foreign organizations for transfer of nuclear technology. Eletrobras finances nuclear construction and advises on the granting of construction and operation permits. The electrical utilities commission construction contracts for and will operate the reactors.[22] The other exceptional case is Iran, the only nuclear dependent country whose atomic energy organization took the lead in directing the country's nuclear effort, and was able to avoid ceding extensive responsibility to utilities by commissioning foreign contractors to implement the entire nuclear program.

In the nonnuclear countries, utility support remains important in gaining governmental approval for a nuclear program. In Indonesia, the state electrical enterprise, PLN, supports nuclear power as part of its mid-1990s expansion program. As in Argentina and Iran, however, the utility's priorities differ from those of the atomic energy commission. Initially, PLN's desire to maintain exclusive control over electricity generation led it to oppose BATAN leadership in nuclear power. In 1972, BATAN officials attempted to assure their place in any future nuclear power program by establishing a Joint Preparatory Committee for nuclear power with PLN. Disagreement also arose over the type of reactor project to select. Baiquni wanted his scientists to gain experience through intimate involvement in the construction of any nuclear reactor. PLN officials wanted a reactor to be built quickly, cheaply, and well, and so preferred a turnkey project to be built by foreign firms. BATAN, understandably restless to build its first power reactor, commissioned studies which showed a 600 MWe unit to be economically feasible. PLN studies, though, judged 1,300 MWe to be the minimum economic size (because the high fixed cost component would be spread over a larger output, installed cost per kilowatt cost would fall). Though this debate seems arcane, its outcome determines when the first nuclear station will be built: the larger the minimal economic size, the longer it will be before installed grid capacity can safely take the additional load furnished by the reactor. In light of their divergent interests, mirrored by diverging analyses, it is not surprising that relations sometimes become strained between PLN and BATAN. Still, PLN will remain a needed ally for BATAN, which must hope that this dependence becomes reciprocal. Despite apparent discord, a high PLN official who moved into the Mines and Energy Ministry, A. Arismunandar, suggested an underlying consensus, when he called the decision to build a nuclear power station the 'political will' of all Indonesians 'so that in all sectors of advanced technology we are not absolutely dependent upon outside sources'.[23]

Central planning organizations, whose responsibilities include paring expenditures and balancing development priorities, arbitrate both between and within ministries. Planners, sharing many inclinations with economic or financial officials – desire to minimize capital outlays, maximize employment opportunities, reduce current account deficits, and avoid uncertain projects – can cause serious problems for nuclear proponents, provided they enjoy strong leadership support. In that proviso they resemble the atomic energy commissions they often oppose, for planners lack the line responsibilities which afford institutional bases for support to their ministerial colleagues. Thus, the Planning and Budget Organization (PBO) in Iran did not seriously threaten the AEOI, which enjoyed prime ministerial support and the Shah's personal protection. Once Hoveyda resigned and the Shah became so preoccupied with survival that he had no time to consider, let alone guard, the nuclear program, AEOI bureaucratic immunity to the PBO and others waned.

Suharto's attitudes toward development and national prestige hurt nuclear power. As noted, he did not identify large-scale nuclear development as essential either economically or politically. His 'New Order' focused instead on the 'technocratic' approach advanced by his close circle of economic advisers, the 'Berkeley Mafia', which included present Mines and Energy Minister Subroto. Their views were given substance in the 1968 reorganization of the state planning agency, BAPPENAS, as a more powerful agency, oriented toward encouraging projects with quicker, more tangible development benefits than nuclear power could offer.[24] BATAN could not escape the state agency's new role as budget clearing house for the government. BAPPENAS became BATAN's nemesis. Its lack of expertise in nuclear technology left the planning agency skeptical, not deferential. Five-year plans always relegated nuclear energy to the same low priority as 'exotic' technologies, such as solar power.[25] The state planning agency would not even support funding of a feasibility study for the first power reactor, so the government did not appropriate the less than $1 million necessary for the project. Instead, BAPPENAS supported expansion of fossil fuel production and the construction of numerous hydropower stations.[26] Fortunately for nuclear advocates, the hopes of BAPPENAS sometimes went unfulfilled. Pertamina posed little threat after its near bankruptcy in 1975, but PN Batubara, the state coal enterprise, was given high priority by BAPPENAS in the mid-1970s. Plans materialized for rehabilitation of the Sumatran coal fields, both to furnish exports and to fuel a large power station on Java.[27] The revival of the coal industry would gravely wound nuclear power plans. By 1980, the disappointing results of PN Batubara's early efforts to revive coal production had revived interest in nuclear power. Nevertheless, BAPPENAS remained wary of the nuclear option.

MILITARY INFLUENCE

One other source of influence is the military. The concern here is *not* military efforts, but rather the influence of the military upon civilian efforts. Its role in governing many developing countries suggests that the military may heavily influence programs. No consistent, particularly military influence, however, can be easily discerned. Sometimes officers lobby for an independent nuclear program, but in this they do not differ from other policy-makers. Besides, it is far from clear what comprises the most 'independent' energy program, and how the nuclear option fits into that assessment. Questions arise not only between coal and uranium, but also between natural and enriched uranium reactors, which entail reliance upon foreign supplies of heavy water and enriched uranium, respectively. One might assume a military preference for the natural uranium cycle, which is more conducive to weapons manufacture, but this is not always the case. For example, in 1972, the Argentine Army backed the Westinghouse light water reactor bid, which eventually lost out to the Canadian tender. Unexpected military attitudes sometimes reflect heavy lobbying by the competitors for a contract, sometimes interservice rivalry. In Argentina, the Navy has always controlled the CNEA, a well-funded, prestigious organization which may well provoke the envy of the other branches. Jealousy may thus sometimes shape military attitudes toward nuclear power.

A clearer military role can be seen in those countries where research is emphasized more than power. A commercial power program can be compromised by military influence or its appearance, as when the North Americans cut off nuclear assistance to India and Pakistan. A research program, though, may be less subject to foreign pressures. The Israeli nuclear program was directed through the Defence Ministry from its inception in the early 1950s.[28] The Kahota enrichment effort reportedly was administered through the Pakistan Defence Ministry, not the AEC. The Libyan and Iraqi nuclear programs are both strongly suspected to bear military intent. Libya signed a contract for missile development with Otrag, a West German firm which has offered missiles to developing countries such as Zaire, Saudi Arabia, and Syria, and which has been reported to have secretly allowed for the military uses of these systems.[29] Following the 1979 La Seyne-sur-Mer incident, the Iraqis refused to accept a less militarily applicable substitute for the originally promised weapons-grade, highly enriched uranium. Such developments intensify the military image of these programs.

DILEMMA OF SCALE

Atomic energy commissions in developing countries face a delicate task in optimizing scale. To plan too modestly risks being ignored. To ask for (and receive) more than a commission can use productively risks falsely raising expectations. The dilemma of scale arises from the need to demonstrate significance, despite limited means. This dilemma applies mainly within government circles, since accomplishments could be exaggerated and blunders hidden from the public. Economic officials are more formidable opponents.

One solution to the dilemma of scale is to advocate an ambitious program, on grounds that anticipated growth of demand requires such large expenditures. Atomic energy commissions which adopt this approach face predictable problems. Government deficits climb. Foreigners, hired at better salaries, incite local resentment (as in Iran), especially since their savings and taxes are likely to be sent home. The complexity of the project causes annoying delays and accompanying electricity shortages. In times of recession, this approach succumbs to reduced national income, which both deprives governments of the ability to support large-scale programs, and reduces electricity demand growth projections.

Alternatively, officials can intentionally overstate the goals of a nuclear power plan. Since an eight-reactor program still must begin with the construction of one, an atomic energy commission might achieve the benefits of large scale – a conspicuous commitment which could boost budgetary support and attract qualified personnel – without the costs of overextension. The deceptive nature of pretended large-scale programs hampers their identification. Pakistan could have one: its commission planned for eight reactors and completed one, while picking up some enrichment technology along the way. Former Iranian officials claimed that their 23,000 MWe target should be interpreted as intentional overstatement, that they recognized at the time that the target could not be reached and adjusted the pace of the program accordingly. By contrast, neither Brazil nor South Korea can be considered pretenders; their investment in nuclear backup industry, such as heavy components plants, provides incontrovertible evidence to the contrary.

Yet another solution is to appear conspicuous without advocating large scale. Argentine and Indian nuclear planners have taken this line, proclaiming nuclear power plans that were far more conservative than those of many far less capable developing countries, but which have become cynosures through emphasis upon nuclear independence. This requires a complete fuel cycle, but not a large number of reactors – shrewd policy, since fuel cycle facilities are far less expensive than reactors. The choice of natural uranium reactors obviates the need for the most expensive fuel cycle facility, the uranium enrichment plant. From the early 1950s, Argentinians mined and milled their uranium reserves. CNEA scientists began fuel element production in 1957. In the late 1960s, they built and operated a pilot reprocessing plant. Projected heavy water production would complete the Argentine nuclear fuel cycle.[30] By the mid-1970s, the Indians had attained some fuel cycle capability at almost every stage, including zircalloy production and plutonium reprocessing.

As independence draws nearer, its prospect invigorates nuclear efforts. Success in actually *achieving* independence may be unnecessary. In fact, Indian and Argentine performance in fuel cycle endeavors has often been weak. In Argentina uranium mining and milling proceeded sporadically. Also, both commercial-scale fuel element production and heavy water plants repeatedly have been delayed. The former was scheduled for 1977 completion; by 1981, the completion date had slipped back to 1983. Indian fuel cycle activities have been equally troubled. A heavy water plant at Kota, built with Indian technology, took ten years to build instead of the four planned. Consequently, all subsequent heavy water plants were contracted on a turnkey basis, but troubles continued. In December 1977, a serious explosion set back operation of a Sulzer Brothers plant by perhaps two years. The impetus to the program derives from the *prospect* of achieving independence. Indeed, once independence is achieved, there could be a loss of momentum and need for new inspiration.

Establishing a salient but attainable objective (such as independence) is preferable to overselling a program deliberately by inflating its objectives in numbers of reactors. The latter course can foster disillusionment among governors and governed alike, and redound to the disadvantage of a nuclear program subjected to their vented frustrations.

CONCLUSION

One product of the direct link which often exists between head of state and atomic energy commission is rough equivalence between the priority given nuclear activities and the interest of the political leadership. Due to governments' limited resources, there is little chance for an atomic energy commission to succeed without direct leadership support. The ability and willingness of the finance, economics, and planning ministries to tamper with nuclear power policies over time varies inversely with political support for nuclear power. The implementation, as opposed to the selection, of a nuclear program is less well matched to the level of political support. Failures in policy execution result either from foot-dragging or physical constraints, both of which are beyond executive control. In Iran, Tavanir was accused of foot-dragging on the installation of power lines to and from Bushehr. Meanwhile, reduced ability to pay cash for nuclear stations slowed AEOI reactor ordering, despite the Shah's support of the plan. Individual leadership is critical, but generally succeeds only by the grace of the political leadership. Once the Shah's active support ceased, Etemad's position became untenable. Atomic energy commission leadership can become less dependent upon the executive, as it did in Argentina, if it achieves some stability and a reputation for success.

Bureaucratic organization is an important, but not independent, determinant of the success of a nuclear power program. The strength of a program reflects the active choice of autocratic executives. The influence of organizational structure wanes for fundamentally important decisions, such as whether or not to build nuclear power stations. Bureaucracies can shape the choice faced by the leadership, but organizational advantages can be overridden. When the Shah was frustrated by the Ministry of Water and Power's sluggishness in conducting a feasibility study for nuclear power, he bypassed the study stage and created an organization instructed to bring nuclear power to Iran, as quickly as possible. Powerful opposition to the CNEA was overruled by Argentine presidents. BATAN's lack of such protection made it easier prey for the type of opposition that the AEOI and the CNEA overcame.

Economic circumstance conflicts with the political necessities of nuclear power programs, resulting in the dilemma of scale. Nuclear power depends upon large scale to be economical and politically salient, but developing economies cannot easily support large-scale nuclear commitments. Argentina and India dealt most effectively with scale. Because of their more industrialized economy and educated work force, the governments were able to select a technically more difficult but less expensive keystone of support: independence. This goal exceeds the capabilities of most other developing countries, which are less able to find ways to appear conspicuous without urging a program beyond their means.

NOTES

1   G. Allison, *The Essence of Decision* (Boston: Little-Brown, 1971), represents the clearest exposition of this line of reasoning.
2   *Business India*, reported by Anil Agarwal in *Nature*, vol. 279, (June 7, 1979), p. 469.
3   H. J. Babha, 'Science and the problem of development', *Science*, vol. 151, (February 4, 1966), p. 545.
4   *New York Times*, November 14, 1978.
5   L. T. Logarta, 'Hell all the way from Bataan and back', *Who* (Philippines), January 3, 1981, pp. 8–10.
6   NRC issuance of an export license hinged upon whether US environmental laws applied to US equipment exports, an issue complicated by the presence of American citizens at Subic Bay naval base and Clark air field. See *Philippine Daily Express*, May 8, 1980, p. 6, and *Bulletin Today* (Manila), May 8, 1980.
7   L. Scheinman, *Atomic Energy Policy in France under the Fourth Republic* (Princeton: Princeton University Press, 1965), p. 210.
8   Republic of Indonesia, Institute for Atomic Energy, *Atoms for Peace in Indonesia* (n.p., 1963), p. 3.
9   BATAN, *Laporan tahun 1970–71* (Jakarta: BATAN, 1971), p. 46. Dollars converted from 110 million rupiah at 1970 rates.
10 A. O. Hirschman, *Development Projects Observed* (Washington, D.C.: Brookings Institution, 1967), p. 154.
11 N. K. Rao, 'Technology and economic development', in Saswinadi (ed.) *Supplement to the Proceedings: ITB – Industrial and Technological Research* (Bandung: BATAN, 1971), p. 16.
12 *International Herald Tribune*, March 9, 1981.
13 Ministry of Energy, Philippines, *Energy 1980–1989*, pp. 28, 56–64.
14 A. L. Hammond, 'Alcohol: a Brazilian answer to the energy crisis', *Science*, vol. 195, (February 11, 1977), pp. 564–6.
15 J. M. Miccolis, 'Alternative energy technologies in Brazil', in V. Smil and W. E. Knowland (eds), *Energy in the Developing World: The Real Energy Crisis* (Oxford: Oxford University Press, 1980), pp. 256–7.
16 World Bank, *Development 1980*, p. 17.
17 United Nations, *Statistical Yearbook 1965*, p. 351, and *Statistical Yearbook 1977*, p. 388; and J. B. Fox, J. J. Stobbs, D. H. Collier, and J. S. Hobbs, *International Data Collection and Analysis* (Atlanta: Nuclear Assurance Corporation, April 1979), p. Argentina-8.
18 Ministry of Economy, Argentina, *Plan de energía electrica*, p. resumen-41.
19 *Nucleonics Week*, November 20, 1980, p. 12.
20 Fox, *International Data*, op. cit., p. India-7.
21 J. A. Sabato, 'Atomic energy in Argentina: a case history', *World Development*, vol. 1 (August 1973), p. 32.
22 Fox *et al., International Data*, op. cit., pp. Brazil-33-5.
23 Quoted in *Kompas*, trans. Foreign Broadcast International Service, October 18, 1979, p. 13.

24 See J. Steven Hoadley, 'The politics of development planning agencies: the evolution of Indonesia's Bappenas', *Asia Quarterly*, no. 1 (1978), pp. 67–78.

25 Nuclear power and uranium were not even mentioned in Republic of Indonesia, Embassy of the Republic of Indonesia, Washington, *Indonesia Develops: Five-Year Development Plan, April 1969–April 1974*, 1970 reprint (Washington: n.p., 1970), pp. 7, 15; and in *Second Five-Year Development Plan (April 1974–March 1979): A Brief Description*, pp. 12–13. *Repelita III* only urged that the nuclear option not be overlooked.

26 R. F. Ichord, Jr., *Indonesia*, in G. J. Mangone (ed.), *Energy Policies of the World* (New York: Elsevier, 1977), pp. 41–2.

27 *Indonesian News*, July 1977, pp. 15–16; and *Electrical Review*, April 14, 1978, p. 10. Technical improvements began to make practical the use of coal which previously was considered too low in heat content for commercialization.

28 Fuad Jabber, *Israel and Nuclear Weapons* (London: Chatto & Windus, 1971), pp. 17–19.

29 BBC, 'Panorama', June 1, 1981.

30 For status of and CNEA perspective on fuel cycle activities, see 'Argentina', *IAEA Bulletin*, vol. 14, no. 6 (1972), pp. 2–9. For heavy water and fuel element production plans, see 'Instalará Argentina la primer planta de agua pesada en Sudamérica', *La Opinión*, April 29, 1976; 'Planta para producir combustible nuclear', *Clarin*, May 17, 1974, p. 11; US Department of State, *Telegram Buenos Aires 7027*, August 28, 1979.

10 Foreign Influences

The final issue to be addressed is how deeply developing country nuclear policies are influenced by foreign intervention. Foreign governments, corporations, and organizations may seek to participate directly in the nuclear policy-making processes in the Third World. Less controllably, the advanced nations influence the developing nations through the examples set by their own energy strategies. This chapter concludes that foreign influence has only limited significance in the decision whether to accept ot reject nuclear power. Foreign pressures can become extremely effective, but only after a pronuclear commitment is made.

DIRECT INVOLVEMENT

Many early contacts between prospective partners in nuclear development occur between governments. Governments of the developed countries communicate with developing countries in several ways. One is the unilateral policy statement. In the early 1970s, reactor bids began to be 'sweetened' by the inclusion of sensitive technologies. West Germany sold Brazil a reactor package which promised to include uranium enrichment and plutonium reprocessing plants. France contracted to sell reprocessing facilities to Pakistan and South Korea. Concern grew that contracts would be won increasingly by the vendor with the 'sexiest' package, offering more and more sensitive technologies, which could be abused for military purposes. The US government opposed the export of sensitive technologies, and pressed the Germans and French to curb this practice. By 1976, both of these governments had unilaterally pledged to refrain from future exports of reprocessing facilities, at least temporarily. Neither has since reneged.

In response to the use of a Canadian research reactor, CIRUS, as a source of the fissile material used in the 1974 Indian test, Ottawa unilaterally conditioned continued cooperation upon stricter safeguards against militant uses of any of its nuclear assistance. After an initial grace period and two extensions lapsed, the Atomic Energy

Control Board put the ban into effect on January 1, 1977.[1] Subsequently, both Pakistani and Indian programs suffered. The Australian government unilaterally refused to begin exporting uranium, pending the conclusions of a panel headed by Justice Russel W. Fox *inter alia* on the impact of exports from the Ranger uranium mine on nuclear weapons proliferation. In 1977, the Second Ranger Report concluded that great economic benefit could result from uranium exports, but that a strict safeguard policy could lead Australia to refuse to supply some countries, including some developing countries.[2] Subsequently, Prime Minister Fraser terminated the uranium embargo.

The most important unilateral policies came from the United States, which remains the world's foremost nuclear supplier. In April 1977, President Jimmy Carter submitted to the Congress a bill which threatened to curtail all nuclear assistance to nations which did not place all their nuclear activities under IAEA safeguards, or which refused to renegotiate existing agreements in order to give the US government greater control over the uses to which its nuclear equipment, technology, and fuel exports might be put. The resulting Nuclear Non-Proliferation Act of 1978 (NNPA) gave countries two years (with possible extensions) to accede to these conditions.[3] The implicit message, that the United States mistrusted the nuclear intentions of all other nations, struck numerous governments as arrogant. This impression was unfortunate, since Titles I and V sought to assist developing countries in developing energy alternatives to oil, and assure all nations that the United States would provide nuclear fuel supplies adequate for their needs. These provisions, however, were widely overlooked and therefore could not salve others' resentment. The Carter policy elicited rebuttal rather than acceptance, since it highlighted developing country economic dependence upon foreign investment and technology. Even apart from the nuclear issue, many governments share the ambivalence toward foreign investment felt in Argentina, where it 'has been viewed alternately as a stimulus to national growth and a form of economic colonialism'.[4] The acute sensitivity of developing country governments to allegedly discriminatory nuclear export policies reflects the completeness of their dependence.

Multilateral démarches to the developing world are also greeted with suspicion. Nuclear supplier nations usually resort to multilateral initiatives in order to present a common front to recipients of nuclear technology. Unavoidably, the developed world appears to be 'ganging up' on the developing countries in order to perpetuate their inferiority. The Nuclear Suppliers Group aggravated this sentiment. Suppliers prefer to communicate by agreement rather than by diktat, but have no alternative to one-way channels when the policies to be expressed either restrict supplies or otherwise offend recipient governments, which invariably deplore and try to circumvent such approaches. For instance, the Argentine government cannily played the Canadians and the Germans against each other until it sundered an agreement by each side not to undercut the other's safeguards in order to obtain the Atucha-2 and associated heavy water contracts.

Bilateral approaches are responsible for most cooperative relationships between developed and developing countries. Agreements for cooperation provide the framework for the type of assistance offered, be it for research or power uses or both. By 1976, the seven charter members of the NSG had bilateral agreements for cooperation with thirty-five other nations, thirteen of which were from the developing world. Fissile materials, technical assistance, equipment and manpower training have been provided. Though promotional, bilateral agreements also include regulatory elements. From the beginning, nations advanced in nuclear science included special clauses in bilateral cooperation agreements designed to verify that the recipient countries used the delivered fissile material only for peaceful purposes. In the early 1960s, these advanced nations (most notably the United States) started to transfer the administration of bilateral safeguards to the IAEA.[5] After the 1974 Indian nuclear test, supplier nations began to view nuclear exports more restrictively, reasserting their rights to impose bilateral controls on nuclear technology transfers. The United States and Canada require the stiffest additional assurances, beyond IAEA requirements, that their exports not be used militarily. Increased restrictions damage bilateral nuclear (and sometimes overall) relationships.

Once governments set the framework, commercial contracts precisely define the shape of cooperation. Vendors vary in aggressiveness, a result of their differences in organization, size of domestic market, capacity, government support, and so on. Countries like France and West Germany, whose reactor industries commenced long after those in the United States – in fact, both Framatome and Kraftwerk Union began as Westinghouse licensees – deliberately built more production capacity than their domestic markets required, in order to exploit economies of scale and growing export possibilities. This built-in pressure to export intensified as reduced domestic demand for nuclear power left reactor manufacturers, particularly in the United States and West Germany, struggling for survival. As one German analyst explained:

Large amounts of capital have been invested in the construction of nuclear power stations by nuclear industries in the industrially developed countries. For this capital expenditure to be amortized a minimum number of reactor units must find a market each year. Because of limited domestic markets, export markets must be found for a certain share of the total number of units produced. However the only potential importers are the developing countries. This is because the major Western European countries generating nuclear power – the Federal Republic of Germany, the United Kingdom, France, Italy, Sweden, Switzerland, and Belgium – as well as Taiwan – prefer to purchase products from their own national firms.[6]

Overall commercial relations between nations naturally influence patterns of nuclear cooperation. West Germany has had a long tradition of friendly relations with Argentina. In the electrical sector, the preeminence of Siemens AG (and its subsidiary Siemens Argentina SA) dates to contacts initiated in the early twentieth century. Siemens' aggressiveness in the Atucha-1 competition showed that its position was well earned. It lined up credit to finance 100 percent of the station's cost, through a $43.8 million loan from the semipublic Credit Bank for Reconstruction (or KFW, for Kreditanstalt für Wiederaufbau) and a $26.2 million line of credit directly from its own resources, repayable at the extremely attractive annual rates of 4.5 and 6.0 percent, respectively.[7] The Bonn government has steadfastly assisted Siemens and others by guaranteeing concessionary finance terms and approving export licenses. Their reputation for reliability served the Germans well in their quests to win nuclear reactor contracts in Argentina.

Equally, overall political relations shape nuclear cooperation. The United States has sometimes suffered for this. For instance, as early as 1959, Argentine President Frondizi visited Washington and discussed a possible trade of raw for enriched uranium during his 1959 visit to Washington.[8] Under the Atoms for Peace program, the US government sent Argentina $462,000 and a large amount of special nuclear material, and trained 220 Argentinians.[9] These good relations subsequently disintegrated, and American manufacturers lost the Atucha-1 contract competition to Siemens. US-Argentine relations generally had worsened by that time, with no US ambassador assigned to Buenos Aires and the likely prospect of further cuts in the trickle of military aid to the country. Tension increased over the supply of heavy water for Atucha-1. The United States was the expected source, but Washington officials were only apprised of this very late in the negotiations and indicated some reluctance to cooperate. This angered the CNEA, perhaps tipping the balance decisively in favor of the German firm. By the time the United States agreed to sell Argentina the heavy water, the damage had been done. the thirty-year agreement for cooperation concluded in 1969 between the two American nations

Table 10.1 *Principal Thermal Reactor Suppliers, by Country*

A. Wholly-owned divisions of private enterprises:

Country	Parent firm	Main business of parent firm	Reactor type	Source of technology	Nuclear generating[c] capacity ordered and installed at home and abroad (March 1979)
USA	General Electric (GE)	Electrical engineering	Boiling water reactor (BWR)	In-house[b]	76.5 GWe
	Westinghouse	Electrical engineering	Pressurized water reactor (PWR)	In-house[b]	96.7 GWe
	Babcock & Wilcox	Heavy engineering	PWR	In-house[b]	27.8 GWe
	Combustion Engineering	Heavy engineering	PWR	In-house[b]	34.1 GWe
FR Germany[a]	Siemens (Kraftwerk Union – KWU)	Electrical engineering	PWR	Former Westinghouse licensee	31.4 GWe
Japan	Mitsubishi	Heavy engineering	PWR	Westinghouse licensee	8.0 GWe
	Toshiba	Electrical engineering	BWR	GE licensee	4.4 GWe
	Hitachi	Electrical engineering	BWR	GE licensee	2.3 GWe

B. Mixed ownership consortia/State owned contractors:

Country	Consortium/ Contractor	Ownership	Reactor type	Source of technology	Nuclear generating[c] capacity ordered and installed at home and abroad (March 1979)
France	Framatome	15% Westinghouse 55% Creusot-Loire 30% CEA (Govt)	PWR	Westinghouse licensee	39.1 GWe (PWR only)
UK	Nuclear Power Corporation	30% General Electric Company 35% UK Atomic Energy Authority (Govt) 35% British Nuclear Associates (mainly construction firms)	AGR	In-house (UKAEA)	10.6 GWe (AGR only)
Sweden	ASEA-ATOM	50% ASEA 50% Govt	BWR	In-house	8.1 GWe
Canada	AECL	100% Federal Govt	HWR (CANDU)	In-house	17.5 GWe

[a]KWU has also constructed BWRs under license to GE, and has supplied HWRs to export markets. The German-Swiss reactor supplier, Babcock-Brown Boveri, which is presently constructing one reactor in FR Germany, under license from Babcock & Wilcox, was not considered in this report.

[b]Technology derived, directly or indirectly, from private involvement in nuclear submarine program.

Source: M. Lönnroth and W. Walker, *The Viability of the Civil Nuclear Industry* (London: Royal Institute of International Affairs, 1979), pp. 14–15.

did not salvage the situation. Indeed, it worsened. In 1975, the United States charged that Argentinians had removed reactor waste containing unseparated plutonium from Atucha-1 without IAEA knowledge.[10] The Carter Administration was unpopular with Argentine officials not only because of its human rights policy and arms cutoff, but also because it instigated disagreeable nonproliferation policies.

The major vendors of nuclear reactors and components and their customers are listed in Table 10.1. All seek to provide moral and material support to potential clients. Vendor involvement ranges from the sale of equipment, technology, and fuel to the far more extensive execution of 'super-turnkey' projects that supply not only the energy facilities but also the installation of access roads, harbors, housing and essential community services. Vendors may train manpower to operate and maintain reactors. The Bushehr contracts were super-turnkey. Architect–engineering firms, such as Bechtel and Ebasco, also promote nuclear power, often serving as consultants to agencies or commissions in developing countries.

Particularly important is the vendor's role in obtaining finance and credit facilities. Without this assistance, most developing nations could not even enter the nuclear market. When asked how to prevent the construction of a French reprocessing plant in Pakistan, one analyst replied, 'Ask for a down payment.' A single reactor with fuel may cost around $1.5 billion, or up to $1.9 billion in the case of the Westinghouse contract with the Philippines. In many countries this would represent a large share of available foreign exchange reserves, as shown in Table 10.2, though of course the payments would be spread over several years. For many, investment in a single commercial-scale reactor would divert a significant fraction of available investment resources, while aggravating serious foreign debt problems.

Table 10.2   *Foreign Exchange Reserves, Third Quarter 1981, Selected Countries (in millions of US$)*

	$		$
Argentina	3,142	Mexico	2,831[a]
Brazil	4,759	Nigeria	5,623
Chile	3,295	Pakistan	752
Egypt	1,153	Philippines	1,999
India	4,086	South Africa	160
Indonesia	5,085	Thailand	1,372
Israel	3,370	Turkey	1,476
Korea, South	2,580	Yugoslavia	1,435

[a]First Quarter, 1981.

Source: *International Financial Statistics*, vol. 35, no. 2 (February 1982), *passim*.

Thus the importance of nuclear supplier financial assistance. A brief examination of the major suppliers reveals the alacrity with which concessional terms have been offered. Authorizations for nuclear power plants and training centers from the US Export–Import Bank, through March 31, 1979, reached $7.25 billion in export value, seventy-six loans worth $4.2 billion, and thirty-eight financial guarantees worth $1.84 billion. For the direct loans, total repayment period has been about twenty years, with no principal repayment during reactor construction (which lengthened to eight or nine years by the mid-1970s). Of the sixty-three loans granted by the time President Carter entered office, the majority (thirty-seven) had a 6 percent interest rate.[12]

The French government routinely finances up to 85 percent of goods and services to be supplied in a nuclear export deal, and may finance the entire project. Its blended long-term interest rate has varied between 6.3 and 7.2 percent for a maximum term of fifteen years.[13] German officials in the Economic Ministry, by contrast, claim that KWU operates strictly by the laws of the free market. In truth, although the German nuclear industry may be less sheltered than its French counterpart, it does not come close to perfect competition. Up to 85 percent of an export can be financed on German credit. The government sometimes offers Hermes guarantees against the loss of foreign investments due to political reasons (such as expropriations or the advent of an Islamic Republic) in order to ease industry's risk. For Atucha-1, the German government offered the Argentinians a five-year, no interest loan, followed by a low interest loan, in addition to balance of payments considerations.

The role of the 'Big Three' banks – the Deutsche Bank, Dresdner Bank, and Commerzbank – is critical in West Germany. They are so large and diversified (holding more than 25 percent of voting capital in twenty-eight of the one hundred largest national enterprises) that they can afford to take larger risks on larger sums than can normal commercial banks.[14] One incentive for them to do so in the nuclear field is to maintain German leadership in the export of high technologies. Another is the general desire to export capital to developing nations, in order to foster economic growth in potential markets for increased German exports, while forging links to secure access to valuable raw materials found in the South. West Germany has negligible uranium resources of its own, and Brazilian uranium imports could prove attractive should the German nuclear program survive its present depression. Aside from international trade advantages, German capital exports could help maintain domestic economic stability. Large trade surpluses increase domestic liquidity, threatening to over-stimulate demand and so fuel inflation. Though other nuclear

Table 10.3 Export–Import Bank of the United States: Authorizations for Nuclear Power Plants and Training Centers from Inception through March 31, 1979

Fiscal Years	No. of Countries	Export Value ($ thousands)			Number of Direct Loans	Direct Loan Authorizations ($ thousands)			Number of Financial Guarantees	Financial Guarantee Authorizations ($ thousands)		
		Equipment	Fuel	Total		Equipment	Fuel	Total		Equipment	Fuel	Total
Gross Authorizations:												
1959	4	122,820	12,530	135,350	4	122,820	12,530	135,350	—	—	—	—
1960	1	26,682	7,318	34,000	1	26,682	7,318	34,000	—	—	—	—
1964	1	2,000	—	2,000	1	2,000	—	2,000	—	—	—	—
1965	1	22,085	5,500	27,585	1	19,000	5,500	24,500	—	—	—	—
1966	1	33,250	10,000	43,250	1	30,034	10,000	40,034	—	—	—	—
1967	4	115,717	56,317	172,034	6	96,955	48,368	145,323	—	—	—	—
1968	2	763	8,750	9,513	2	763	8,750	9,513	—	—	—	—
1969	2	102,024	34,967	136,991	3	87,322	27,974	115,296	—	—	—	—
1970	4	179,836	81,044	260,880	7	127,414	46,351	173,765	3	25,131	9,429	34,560
1971	5	451,410	144,417	595,827	10	211,901	56,191	268,092	10	193,162	69,027	262,189
1972	4	985,327	224,019	1,209,346	7	652,097	72,946	725,043	5	224,312	74,971	299,283

1973	4	272,427	57,924	330,351	5	141,093	34,333	175,426	4	138,321	9,133	147,454
1974	5	713,424	60,669	774,093	6	403,611	35,740	439,351	5	147,506	14,940	162,446
1975	7	374,420	109,211	483,631	7	162,879	41,923	204,802	3	123,371	22,734	146,105
1976	4	1,261,992	197,292	1,459,284	6	566,221	81,509	647,730	5	519,008	52,985	571,993
TQ	1	144,600	16,100	160,700	1	73,370	8,096	81,466	1	46,284	5,250	51,534
1977	1	137,000	26,000	163,000	2	75,350	13,100	88,450	1	46,959	7,291	54,250
1978	4	828,381	421,398	1,249,779	5	618,734	268,980	887,714	2	79,609	27,000	106,609
1st half 1979	–	–			–	–			–	–		
Subtotal	75	5,774,158	1,473,456	7,247,614	75	3,418,246	779,609	4,197,855	39	1,543,663	292,760	1,836,423
Net Authorizations[a]	76	5,763,144	1,473,456	7,236,600	76	3,386,200	797,140	4,183,340	38	1,446,360	269,970	1,716,330

[a]Adjusted for prior fiscal year reversals and transfers, guarantee repurchases, and supplier participation which was included in gross authorizations.
Source: Congressional Research Service, Library of Congress, *Nuclear Proliferation Factbook* (Washington: US Government Printing Office, 1980), p. 301.

suppliers may not share West Germany's trade surplus 'problem', they also would like to nourish foreign markets for the sake of securing future export buyers and raw material suppliers. Nuclear reactor exports also contribute to domestic employment, which appeals to all suppliers.

One might expect reactor exports to be lucrative. They may not be. Since contracts are so large and few in number, bidders are willing to bargain hard for the privilege of gaining a toehold in other nations' energy markets. They have even been willing to sell reactors as loss leaders when necessary. As stated in the Barber study, the 'German, US, and Canadian vendors "lost their shirts" on their initial sales to Argentina, India, and Pakistan [and] sold reactor systems which even with the subsidized terms were not economic'.[15] Consequently, neither governmental nor industrial supplier willingness to offer favorable prices and terms is absolute. As early as 1974, Westinghouse changed its pricing policy regarding major turnkey power plant projects overseas to reflect the rapidly escalating cost of domestic and foreign materials and services. The earlier policy of selling reactors on a firm price basis, to facilitate financing determinations, was abandoned in order to include cost escalation.[16] The large Western electrical manufacturers, such as Westinghouse, General Electric, Siemens, and Creusot-Loire (70 percent owner of Framatome), could afford to be choosy. Their reactor activities comprise only about one-tenth of total power production exports. In the 1970s, exported reactors accounted for around 18 percent of total nuclear capacity ordered, in both North and South. Two-thirds of the 46 GW in nuclear reactor orders in the decade were placed by 1974. Some have already been cancelled (Iran) or are in jeopardy (Brazil).[17]

The strength of promotional foreign influences therefore depends on the capabilities and willingness of supplier nations. Countries such as Belgium, Italy, and Japan expressed interest in selling nuclear reactors to Iran, but lacked the capability to compete with West Germany, France, and the United States. Relative capabilities shifted over time. Manufacturers in France and West Germany were catching up technologically with the Americans by the early 1970s, and consequently exerted stronger influence abroad. Nevertheless, West German influence weakened between the competitions for the Atucha-1 and Embalse contracts, because by 1972 KWU could not offer a single unit heavy water reactor large enough (600 MWe) for CNEA specifications.

Willingness to export also fluctuated. The United States was the world's leading promoter from 1954 until the early 1970s, when apparent willingness to sell slipped markedly (against American manufacturers' wishes), as the government (misguidedly expecting a shortage of capacity) closed the order books for US enrichment services, and toughened its nonproliferation policy. The Canadian government and vendors vacillated. Their initial eagerness was dampened first by India's use of a Canadian-supplied reactor to acquire the plutonium used in its nuclear test and then by the heavy financial losses suffered on the Embalse project, but has since revived. The British have always been world leaders in nuclear technology, but have never exploited this advantage through an energetic export drive. The French and West Germans became the most willing suppliers among the advanced nations.

The success of the various approaches taken by vendors depends upon the perceptions of potential customers, whose concerns fall into three important categories: technology, price, and reliability. First, what technology does each manufacturer offer: light or heavy water reactors, or fuel cycle technology helpful to achieving nuclear self-sufficiency? Second, which offers the best combination of price and financing terms for a good product? Third, which vendors seem most likely to fulfill all commitments, offering what guarantees for fuel supply and against nonperformance?

Clearly, the attraction of international nuclear trade is not unmitigated for either buyer or seller. Consequently, each holds leverage over the other. Whichever side is prepared to retire from negotiations without a contract enjoys a bargaining advantage over the more eager side. The Argentinians, South Koreans, and Taiwanese are able to reduce reactor price tags because all are seriously committed to nuclear power and can therefore allow vendors to battle for the best offer. Once a commitment is made by an importer, however, some advantage shifts over to the exporter, who has now obtained a captive market and so can exert leverage by withholding, delaying, or threatening supplies of reactors, fuel services, and fuel. For instance, in the renegotiation required by the Philippine reactor cost escalations following President Marcos' suspension of construction, Westinghouse was able to increase the price by over 70 percent, from $1.1 billion to $1.9 billion. The initial Westinghouse bid offered *two* 600 MWe units for $1.2 billion. Remarkably, the Philippine government even waived contractor liability under Article 1715 of the Civil Code of the Philippines, which requires the vendor: 'to execute the work in such a manner that it has all qualities agreed upon and has no defects which destroy or lessen its value or fitness for its ordinary or stipulated use'.[18] This left the Filipinos strait-jacketed in negotiations over sharing the cost increases involved in modifying the reactor in the wake of the Puno Commission report. Reflecting its dominant negotiating position, just after arriving in the Philippines, the six-man Westinghouse negotiating team took umbrage, walked out of the talks, and returned home.[19]

Irritating supplier behavior of course tarnishes the image of reliability, which can be decisive in a developing country government's selection of trading partners. Argentine-Canadian cooperation provides a good example. The metamorphosis of the financing arrangements for the Embalse reactor began with a reported price of between $220 million and $250 million, of which $100 million (roughly 40 percent) was to be financed through the Export Development Corporation of Canada.[20] Canadian eagerness and ineptitude, plus Argentine shrewdness and an economy careening through massive inflation and currency devaluations, rapidly pushed up the price, costing AECL President John Foster his job and his firm hundreds of millions of dollars. AECL twice demanded contract renegotiation. Twice the CNEA reluctantly agreed. The delays caused by the first renegotiation alone cost $129 million, in Castro Madero's estimation, raising the total cost to $800 million when added to the other cost overruns.[21] The Embalse price by 1981 had reached $1.2 billion. Far from moving to the rescue, the Canadian government further aggravated the situation by insisting that Argentina accept tighter safeguards, such as a plutonium ban.[22] Canadian behavior in the Embalse project biased the CNEA against the AECL in the Atucha-2 negotiations.

DOMESTIC EXAMPLES: INFLUENCE BY CIRCUMSTANCE

So far intentional influences have been discussed. Unintentionally, however, the developed countries influence developing countries through their own, more advanced, nuclear power programs. In the late 1950s and early 1960s, commercial nuclear power appeared ready for widespread application, as reactors began supplying electricity to American, British, and French consumers. Large investments in the advanced countries reflected growing confidence in nuclear power. By the mid-1970s, projections that tens to hundreds of thousands of megawatts would be needed in numerous countries were replaced by lower forecasts, due to recession and the increasing unpopularity of nuclear power. Light water reactors failed to stabilize in price, raising doubts about their long-term viability. From reports of near accidents and radiation leaks, culminating in the 1979 Three Mile Island (TMI) incident, concern grew that nuclear power and the storage of its wastes presented serious health hazards. Debates over nuclear power erupted between governors and governed, as well as within ministries, political parties, trade unions, and scientific academies. Orders for nuclear stations slowed to a trickle while cancellations of existing orders increased (except in France and the Eastern bloc). In West Germany, domestic reactor orders ceased after June 1975, and KWU Chairman Klaus Barthelt claimed three years later that the West German nuclear industry would be bankrupt without overseas orders.[23] The scaling back of the French nuclear program under President Mitterand cast into doubt the future of the most untrammeled nuclear program outside the Eastern bloc.

Officials in developing countries have watched this evolution carefully. Atomic energy commissions have defended; finance and health ministries have questioned. Two divergent effects have emerged. First, many in developing countries no longer view nuclear energy as the key to closing the technological gap. Slowing Western programs have afforded authorities in developing countries time to pause for reappraisal. The capital costs seem increasingly oppressive as these nations struggled with recession, inflation, and underemployment. The second effect has been the redoubled ambition of reactor manufacturers to find outlets for their undertaxed capacities. Competition for the developing country market has intensified. In sum, the situation in the advanced nations could impel developing nations to or from commitments to nuclear power.

Given the more relaxed attitude in the South toward environmental protection, perhaps the most searing impression of the TMI incident was its enormous financial cost. General Public Utilities, owner of the unfortunate unit, nearly went bankrupt purchasing substitute electrical supplies from neighboring utilities and paying other penalties for its own cutback of generating capacity. Were insurance premiums fully internalized, they surely would have risen, implying that the incident increased the hidden costs, or externalities, of nuclear power. Moreover, a $1 billion investment became, for an indeterminate time, a white elephant. What developing country could sustain a similar loss without drastic economic repercussions?

The three case studies discussed in Part Two reveal the disparate effects that domestic examples in the advanced nations can have in the developing world. At the 1950 inception of the Argentine nuclear energy program, the experience of other nations was too limited to offer much guidance. Commercial nuclear power was several years away in even the most advanced nations. The country which later would supply the Atucha-1 and -2 reactors would not obtain Allied permission to engage in its own nuclear activities for five more years. Thus the early Argentine path was largely self-determined. By the time that doubts about nuclear power slowed programs elsewhere, the CNEA commitment was too deeply entrenched to be dislodged easily. Refusing to be discouraged by radioactivity leaks and reactor shutdowns in West Germany and other advanced nations, the CNEA proudly pointed to Atucha-1 as the reactor with the highest operating capacity (the percentage of time that the reactor was operating, compared to the percentage during which it was shut down) in the world. Throughout, Argentine debate never focused upon *whether* nuclear power was desirable, but rather on *how* it could best satisfy Argentine needs.

In Iran, the example of the nuclear programs in the advanced countries had an erratic effect. Through the early 1970s, the only product of the Shah's interest in nuclear energy, necessarily a product of foreign example, were tiny research efforts set up at Tehran University under CENTO and the Atoms for Peace program. Perhaps the progress of nuclear power in the developed world heightened the Shah's frustration at the lack of progress in Iran, creating a sense of lost time which led him to propose such a large program. To execute its task, the AEOI could not tarry, and had to be impervious to the crisis of nuclear power in advanced countries. Accumulated nuclear power station experience had satisfied the Shah of the value of fission. It was now time to act. AEOI officials saw the flaw of nuclear power programs in the developed nations not in inadequate security or radiological protection at reactors, but rather in the tendency to become mired in a morass of burgeoning regulations and ill-informed public protests.

The domestic examples in advanced nations must have served Indonesian nuclear planners as distant aspirations, not direct influence. These examples certainly awakened Indonesian interest: no nation so financially impoverished yet naturally blessed would have decided on its own to seek to exploit the one major energy technology whose fuel it was not known to possess in significant quantities. The impact was not deep, however, and when the prevailing winds shifted against nuclear power, there were no interests sufficiently entrenched to maintain a serious power program.

OTHER ACTORS

Atomic energy is of such compelling military, economic, social, and environmental interest that it is not surprising that, in the words of Jimmy Durante, 'Everybody wants to get into the act'. Advocates and antagonists of nuclear power are arrayed against each other, and one cannot yet be certain which side will prevail. The nuclear and electrical utilities industries have large and impressive resources to substantiate their case. Often they can count on governments to act on their behalf, at least once the political leadership has staked its energy policy (and so its general reputation) on the nuclear option. Some lobbying organizations are transnational or are formed as a number of individual but communicating national entities, such as Foratom, the Uranium Institute, and the Atomic Industrial Fora.

Nuclear power antagonists are as committed as their opponents, and have been extremely effective in either sensitizing or unduly alarming (depending on one's viewpoint) the public to the dangers of nuclear technology. Even national organizations occasionally intervene in the Third World nuclear scene, as when the American-based Natural Resources Defense Council sought to convince the US Nuclear Regulatory Commission not to issue an export license for the export of a nuclear reactor to the Philippines, because of the seismic and volcanic instability of the site.[24] Philippine opponents to the plant submitted a report by an American geologist to the Puno Commission, supporting their claims. Foreign opposition often finds local sympathy. In the Philippines, it at least partly coincides with anti-Marcos sentiment. In Brazil, the scientific community preferred indigenous nuclear development, which it could dominate, to massive German imports, which it could not. Expatriates and apostates form a final, overlapping category of nuclear critics in the developing world. Robert Pollard, who resigned in protest from the US NRC, participated in the Philippine debate over PNPP-1.

Foreign advice elicits two opposing responses. The first is suspicion. The Argentine decision to break precedent and conduct its own reactor feasibility reflected mistrust of the possible advice of unpredictable outsiders, as well as self-confidence. The second response is respect. Decision-makers in developing countries sometimes give greater credence to the opinions of well-established foreign experts and international organizations than to their own agencies, which may be perceived to be less competent and objective. In the Puno Commission hearings, both sides submitted expert testimony from foreign consultants. Whether suspicion or respect characterizes national attitudes depends upon both the configuration of political forces (for example, the Argentine economic nationalists versus the developmentalists) and the reputation for objectivity enjoyed by the outsider in question.

INTERNATIONAL ORGANIZATIONS

The influence of an international organization on developing country nuclear energy policies depends upon its mandate and disposable funds. Several international organizations have mandates in the nuclear field, including the IAEA, the Nuclear Energy Agency of the Organization for Economic Cooperation and Development, the European Atomic Energy Community (Euratom), the Inter-American

Nuclear Energy Commission of the Organization of American States, the Latin American Energy Organization, and the UN Atomic Energy Commission. Nuclear planners sometimes seek financial support from international institutions with overall development mandates, such as the International Bank for Reconstruction and Development (IBRD, or World Bank), the Asian Development Bank (ADB), and the UN Development Program (UNDP). Some international organizations of broader mandate also take occasional initiatives, as when the European Economic Community Commission drafted proposals for uranium prospecting and the eventual establishment of a nuclear energy program to aid developing countries.[25]

Few of these organizations can offer much money to assist developing country atomic energy commissions. Most serve as occasional discussion fora and little else. The IBRD receives funding requests for projects ranging from hospitals to houses, nutrition to public health, irrigation to road construction, coal to hydro to nuclear power. Its broad mandate is typified in the sentiment expressed in this reply of a former IBRD president to UN Secretary General U Thant in 1969:

> It would do a disservice to our member countries and be a misallocation of development resources, were we to finance a nuclear energy project which did not appear to be both a priority project from the point of view of the economy of the country as a whole and also the most economically advantageous of the various power alternatives available to the country at the same time. Therefore, we believe that the criteria applicable to nuclear energy projects should be the same as those applicable to other kinds of projects coming to the Bank for financing.[26]

Specifically, Bank officials have argued that without national grids or modern appliances to plug into them, it makes no economic sense to build big generators, especially capital-intensive, nuclear-powered ones.[27] As a result, the IBRD, ABD, and IADB have all refused to fund any nuclear projects in developing countries, though they have supported numerous oil, coal, and hydro projects there. The UNDP for a time provided around $4 million per annum for projects executed by the IAEA.[28]

The only international organization with both suitable mandate and funds to back it is the International Atomic Energy Agency. The creation of such an agency was first suggested publicly on December 8, 1953 in US President Dwight Eisenhower's Atom for Peace address before the UN General Assembly. 'A special purpose' of the Agency, he stated, 'would be to provide abundant electrical energy in the power-starved areas of the world'.[29] The statute of the Agency proclaimed the objectives of accelerating and enlarging the contribution of atomic energy to peace, health, and prosperity, while ensuring that assistance provided by or through it would not be used to further any military purpose. The Agency was to take 'due consideration for the needs of the underdeveloped areas of the world'. Between 1958 and 1973, the IAEA provided 4,300 fellowships, the services of over 1,900 experts, and equipment, together valued at $36.6 million.[30]

After 1966, the emphasis of the IAEA gradually shifted from reactor theory to the practical aspects of nuclear power stations. By 1976, nearly one-third of all fellowships and half the training courses were devoted to training manpower in developing countries. In co-operation with West Germany, France, and United States, the Agency ran a series of fifteen-week courses in 1975 and 1976 on nuclear power project planning, implementation, construction, and management. The Agency budget for fellowships increased from $250,000 in 1958, to $2 million per annum in the 1960s, to $6 million in 1977. IAEA Director General Eklund concluded that in the late 1970s, the Agency was annually spending or administering approximately $15 million that contributed 'directly to the transfer of technology to developing countries'.[31] Apart from technical assistance in training, uranium prospecting, and research reactor programs, the IAEA also assisted developing country atomic energy commissions by convening numerous conferences addressed to their needs. In 1960 and 1970, Agency symposia studied the prospect of small and medium power reactors. The foreword of the 1970 report 'pessimistically' forecast that nuclear power would comprise less than 10 percent of installed capacity by 1980, a figure which still proved to be grossly over-optimistic.[32]

Wishful thinking seems inherent to the distorted conclusions of pronuclear fora. The 1973 and 1974 IAEA market surveys for the developing countries graphically illustrate the point in Table 10.4. In its 1977 fuel cycle conference at Salzburg, the IAEA projected installed capacity in developing countries to be 290 to 440 GWe by 2000, or 20 percent of world nuclear capacity. By 1979, members of the West German nuclear research institute at Jülich concluded that by century's end nuclear capacity in the developing world would reach only 83 GWe, or 2.5 percent of world nuclear capacity.[33] The most favorable thing one can say about these estimates is that their purpose is hortatory rather than predictive, and that the IAEA could help legitimize the nuclear option in countries where foreign advice is respected more than suspected. On the other hand, the routine exaggeration which inevitably debases the credibility of any forecast may also taint whoever succumbs to it, including the IAEA.

Table 10.4   *Projected Nuclear Plant Additions by 1990, IAEA Market Surveys (in MW)*[a]

Country	1973 Report			1974 Edition Nuclear total (MW)
	Total nuclear additions	Total thermal market	Nuclear % of total market	
Argentina[b]	6,000	6,800	88.2	6,600
Bangladesh-L		1,300	0	4,000
Bangladesh-H	600	3,850	15.6	
Chile	1,200	1,750	68.6	1,700
Egypt	4,200	4,800	87.5	5,000
Greece	4,200	4,500	93.3	5,000
Jamaica-L		1,000	0	1,750[c]
Jamaica-H	300	1,550	19.3	
Republic of Korea	8,800	9,100	96.7	8,600
Mexico	14,800	19,600	75.6	20,900
Pakistan	600	2,000	30.0	4,800
Philippines	3,800	5,400	70.3	4,800
Singapore-L		2,100	0	4,250
Singapore-H	2,600	4,700	55.3	
Thailand	2,600	3,850	67.5	3,700
Turkey-L	1,200	3,000	40.0	5,000
Turkey-H	3,200	4,850	66.0	
Yugoslavia-L	4,800	6,000	80.0	10,000
Yugoslavia-H	9,200	10,600	86.8	
Total nuclear (L)	52,200	71,200	73.3	86,100
Total nuclear (H)	62,100	83,350	74.5	

[a]L = market survey low load forecast; H = country high load forecast.

[b]Markets for countries with one load forecast are included in both low and high load totals.

[c]Not including the 100 MW unit.

Source: IAEA, *Market Survey for Nuclear Power in Developing Countries: General Report* (Vienna: IAEA, 1973), p. 5; and IAEA, *Market Survey for Nuclear Power in Developing Countries, 1974 Edition* (Vienna: IAEA, 1974), p. 10.

The IAEA has regulatory as well as promotional responsibilities. Safeguards first developed bilaterally, in special clauses to cooperation agreements reserving the right for the donor country to verify that the recipient country did not divert fissile material to military purposes. While the United States began transferring the administration of bilateral safeguards to the IAEA, some members voluntarily submitted nuclear activities to Agency safeguards. The IAEA noted that:

At that time safeguards concerned *only* the material or the nuclear plant which they were intended to safeguard from possible diversion to military uses. It was therefore quite possible for a country to have one reactor under safeguards and just next to it another reactor – built either purely by national means or with the help of a country that did not require safeguards – which could be put to other uses than peaceful ones. [Emphasis original.][34]

For all safeguarded materials, governments must (1) permit IAEA to review the design of existing and planned nuclear plants, (2) submit a precise account of all nuclear material flows, and (3) allow Agency inspectors to verify submitted information.

Sometimes sealed TV cameras are emplaced to take photographs of containers at intervals shorter than required to divert the safeguarded materials. Other cameras commence filming when a container or vault is opened. Also, seals which must be broken to open containers of safeguarded material are inspected. Together, these measures cannot prevent diversion, but safeguards it is hoped will provide 'timely warning' of violations, so that the international community may respond before events overtake them. Once the Agency finds itself unable to verify the use of safeguarded materials, or actually detects a violation, Statute Article 12.C obliges it to report the situation to the UN Security Council.

The NPT sought to reduce the ability to build unsafeguarded facilities through safeguarded assistance. Article III requires that agreements must be concluded to safeguard 'source or special fissionable material in *all* peaceful nuclear activities within the territory' of nonnuclear weapon states party to the Treaty. (Emphasis added.) Non-NPT members remain free to build indigenous, unsafeguarded facilities. To close this loophole, in the mid-1970s, France and West Germany sought to permit export of sensitive facilities to non-NPT members while barring the use of the transferred technology to build unsafeguarded facilities. This 'nonduplication' proviso was first applied to the proposed sale of a reprocessing facility to South Korea and to the Germany–Brazil deal. The nuclear facilities which remain unsafeguarded are listed in Table 7.1. The United States and Canada adopted a more drastic approach, attempting to extend NPT-type safeguards, known as 'full-scope', even to non-NPT members, against the implacable opposition of such countries as Argentina and India.

The IAEA serves different needs as a nuclear program develops. The Argentine example shows how the Agency can be manipulated for various purposes, as the CNEA used it as a source of technical support, a forum to express its views to the world, and, finally, as a foil to help it advertise its growing independence. The government actively supported the IAEA from the outset, providing a chairman to the Board of Governors, organizing and hosting numerous conferences, conducting studies, and providing technical assistance to nations less developed in the nuclear field. Admittedly, international contributions were small. Total IAEA and UNDP technical assistance to *all* Latin America at the beginning of the 1970s was under $1 million annually and only passed $1.5 million in 1975.[35] Nonetheless, this aid was useful. Throughout the 1960s, Agency-provided experts, equipment, and funds assisted the CNEA in such fields as uranium prospecting and production, reactor instrumentation, and medical and agricultural applications of radioisotopes. The IAEA role was important in preparing the 1967 ten-year nuclear plan. The government also signed an agreement for nuclear cooperation with Euratom, which provided for the exchange of information and personnel.[36] As CNEA activities reached the commercial stage, however, the benefits from Agency assistance diminished, to the point where Argentina in 1979 became the second nation (India was first) to withdraw from the Agency's technical assistance program. Castro Madero explained the move as an indication of 'the level of nuclear development we have reached', and also as a protest against the IAEA's overemphasis on safeguards at the expense of technical assistance.[37]

Critics of the IAEA see it as a den for indiscriminate nuclear promotion, which inadequately heeds its regulatory responsibilities. Safeguards abrogations, neither reported nor corrected, have been alleged. The paucity of safeguards inspectors – around 130 in 1981, responsible for over 600 facilities in some fifty countries[38] – is blamed for making safeguards enforcement virtually impossible. Promotion conflicts with regulation, they argue, and the vested interest of the Agency coincides with that of its clients (the member states' atomic energy commissions) who prefer to emphasize the former. As evidence, they note that safeguards costs comprise only a quarter of the regular IAEA budget of $80 billion for 1980, though this had risen from under 19 percent in 1976 and from under 12 percent in 1971.[39]

Israel most brazenly expressed lack of faith in the IAEA system by attacking the safeguarded Iraqi research reactors Tammuz-1 and Tammuz-2 (known before the Camp David accord by the Egyptian names, Isis and Osirak). Prime Minister Begin claimed to have irrefutable evidence that the Iraqis intended to build nuclear weapons for deployment against Israel. Circumstantial evidence strongly supported the allegation. The French had agreed to supply Iraq with a 70 MW research reactor similar to the French Osiris built at the Saclay nuclear research center, to the consternation of many, especially the

Israelis. The Iraqi model was to be fueled by weapons-grade uranium, enriched to 93 percent uranium-235. After the cores were heavily damaged in April 1979 by explosions set by saboteurs at the southern French port of La Seyne-sur-Mer, the Iraqis successfully resisted French pressure to accept a modified reactor which would use less highly enriched and therefore less weapons usable uranium.[40]

Whatever the political and strategic wisdom of the Israeli attack, it constituted a frontal assault upon the IAEA safeguards system. The IAEA all along had stressed that Iraq was an NPT member, and that all the fissile material was fully accounted for and closely monitored. After the attack, Director General Eklund told the IAEA Board of Governors that 'in a reactor of this type, diversion of fuel elements or undeclared plutonium produced at a low rate cannot be technically excluded but would be detected with very high probability', which 'means, in this case, a full guarantee, since the main inspection activity consists in counting about sixty fuel elements which are individual objects of considerable size. Additionally, measurements are made which prove that the elements are not dummies.'[41] Eklund declared publicly that safeguards at the Iraqi nuclear research center containing the research reactors 'have been satisfactorily applied to date, including during the recent period of armed conflict with Iran', rightly concluding that 'it is the Agency's safeguards regime which has also been attacked'.[42] Essentially, because the IAEA can only detect diversions *ex post facto*, and can prevent neither safeguards violations nor withdrawal from the NPT or IAEA safeguards systems, its usefulness ultimately depends on good faith, a commodity palpably lacking in the Middle East. The dependence on good faith cannot be avoided, since application of safeguards requires some cession of sovereignty in the first place, and governments resist further encroachments while remaining able to reassert their sovereignty at will, though commercial contracts impose additional, less voluntary constraints.

The Agency is also criticized by the pronuclear lobby. Atomic energy officials argue that an imbalance in the IAEA role developed in the 1970s, overemphasizing safeguards at the expense of 'other activities the IAEA could perform to aid consumer nations in the enjoyment of the peaceful atom'.[43] This view is common in developing countries, where nuclear promoters remarked that, despite IAEA assistance, not one developing country nuclear power program had ever contributed significantly to total domestic electricity consumption. In 1976, nuclear power comprised only about 2 percent of installed electrical capacity in India and Argentina, the developing countries most advanced in nuclear development. In several developed countries, nuclear power contributed 10 percent or so to electricity production. This relatively poor record helps explain the Indian and

Argentine withdrawal from the Agency's technical assistance program. A similar disgruntlement could be perceived in Iran, where the Agency sent a two-man team to report on nuclear power prospects. The 1973 study concluded that it would be impossible to install a commercial power reactor in Iran before 1983, but that given increased oil prices, a nuclear program should 'from a national economics point of view be pursued vigorously'.[44] The team recommended the establishment of a nuclear power project group in the Ministry of Water and Power which would begin site selection for the smallest commercially available reactor of proven design (500 to 600 MWe), even though projected electrical demand indicated that 800 to 1,000 MWe units could safely be accommodated. No further stations were suggested.

This report confirmed the AEOI view of the IAEA as too languorous and plodding ever to contribute to the establishment of a viable power program. Iranian officials believed that an opportune moment had to be seized boldly. Another might never arise. Studies led down an ever-darkening path to nuclear obscurity, not nuclear power. AEOI officials ignored this report and did not return to the Agency for further assistance in its power program. The Organization needed the IAEA neither for legitimacy, obtained through the Shah, nor assistance, obtained from other countries in return for Iran's new wealth.

Both critiques have merit. Compromise and, as a result, unhappy partisans are inevitable when an organization is charged with both promotional and regulatory responsibilities. From the outset, Agency promotional resources were intended to be extremely limited. Statute Article XI.B provides for a clear IAEA role as broker between financing institutions and member states, but not as guarantor or source of credit. Resources for regulation are even more limited. Though the total IAEA budget increased more than fivefold in the 1970s, the 1980 figure of $80 million (for perspective) equaled less than two-thirds the final cost of the Atucha-1 reactor. Perforce the Agency has to act as a promotional catalyst, not bankroller, and a regulatory spot checker, not sentry.

These roles, though modest, are important. Agreement on a common set of safeguards and inspection procedures established a worldwide regime based upon binding commitments opposed to the military uses of nuclear materials. These commitments were given force by the scope of Agency membership – 110 nations – and by the imposition of IAEA safeguards system, despite its inevitable shortcomings.

Promotionally, for many developing countries, small-scale assistance remains significant to their small programs. Eklund claimed that developing country absorptive capacities, more than shortage of

IAEA assistance, limited technology transfer possibilities.[45] Also, the value of an expert sojourning three months in a developing country can far exceed the nominal value of his time. Finally, moral support and information exchange obtained through the Agency could benefit small atomic energy commissions in their efforts to encourage domestic governmental support.

Apart from permanent international organizations, from time to time other international fora play a role in developing country energy policies. In the 1970s, the United Nations sponsored conferences on such subjects as the environment and science and technology for development. Tangible results, such as the establishment of the UN Environment Program in Nairobi, sometimes arise, but often these exercises have proved feckless. The most important special forum in the nuclear scene to date has been the International Nuclear Fuel Cycle Evaluation, proposed by President Carter as a way to discover and promote methods to prevent the military use of nuclear technology. Stipulating that INFCE would be a nonbinding study and not a negotiation, sixty-six governments and five international organizations took part.[46] While continuing the tradition of wild optimism which seems endemic to gatherings of nuclear advocates, the INFCE final report did acknowledge, without major dissent, that civil technology and facilities 'could be drawn on for a subsequent nuclear weapons program', and noted possible techniques to reduce that prospect. The two-year evaluation, which ended in February 1980, did not make or break any new or existing nuclear policies in the developing world, but it advanced the cause of nonproliferation by defusing the liturgical insistence that nuclear weapons and nuclear energy are entirely distinct and that proliferation is a political *rather* than a technical problem. In fact, it is both. INFCE was considered sufficiently useful as a vehicle for narrowing differences among governments to justify establishment of a Committee for Assurance of Supply (CAS). Since developing country governments recognize that nuclear technology transfers can only take place when enjoying the confidence of nuclear suppliers, they may permit CAS discussions to influence their nuclear policies.

INTERDEVELOPING COUNTRY INFLUENCES

Nations within the developing world influence one another. India has concluded agreements for nuclear cooperation with Argentina, Iran, Libya, and others. Both Brazil and Argentina signed agreements with all major South American governments, and in 1977 Argentina entered the ranks of the nuclear equipment exporters, agreeing to build a 10 MW research reactor in Peru for radioisotope production and scientific training.[47] The CNEA was slated to provide $25 million of the estimated $50 to $70 million cost. (Interestingly, the 315 MWe Atucha-1 cost $70 million.) The Argentinians were pleased to have outbid Canada, Great Britain, Spain, and West Germany. In March 1981, Argentina agreed to perform a nuclear power feasibility study for Uruguay, and may eventually supply a research reactor. Castro Madero projected Argentine nuclear fuel cycle self-sufficiency for 1984, and declared that 'we want to be the exporter of nuclear technology in Latin America'.[48] India is seeking to become a major donor of aid and assistance in the developing world, not only in nuclear power but also in hydro and thermal power stations, in Nepal and Libya, respectively.[49]

Recipients welcome assistance from other developing countries as a substitute for dependence upon the major powers. Also, technology exported from countries like Argentina and India may be more appropriate to other developing countries than are those derived in the alien political, economic, and social environments of the industrialized world. Donors gain prestige for their technical prowess and perhaps improved status within the nonaligned movement. Further, laying a simple groundwork for cooperation today may facilitate an emergent exporter's successful market entry tomorrow, if profits then become possible. Finally, the large scale required to make a domestic nuclear industry economical may catalyze export interest. In Brazil, for example, Nuclep was built on schedule, able to produce heavy components for two reactors each year. The growing delays in the Brazilian reactor program, however, left the new facility idle. Following the May 1980 Argentine-Brazilian nuclear cooperation agreement, Argentina commissioned Nuclep to weld and assemble the lower part of the Atucha-2 pressure vessel, which was to be cast in Japan.

Cooperation does not always characterize interdeveloping country nuclear relationships. The Israelis attacked Osirak, and Indians have occasionally threatened to destroy Pakistani bomb-making efforts. Reportedly, the Saudis sought to replace Libya and Iraq as financial backers to the Pakistan nuclear effort, in order to prevent an 'Islamic bomb' contagion from breaking loose in the region. Brazil and Argentina, despite their 1980 agreement, traditionally have been rivals in the Latin American nuclear field. The refusal of the People's Republic of China to accept IAEA safeguards on its nuclear activities reinforces Indian resistance to the same.

Multilaterally, there has been little tangible result from nuclear relations in the developing world. Various inter-American bodies have held conferences periodically since the 1950s, but without significant issue. Typical were the efforts for Asian cooperation to help small nuclear programs achieve together what none could achieve alone, beginning with a 1955 American offer to furnish $20 million in Atoms for Peace assistance toward the establishment of a regional nuclear research and training center. Even after a trip to Brookhaven National Laboratories by Asian energy officials, the response was only luke-warm, since in most of the South Asian nations 'other claims on the supply of scientific and technical men, financial, and material resources would, at least in the initial phases, of necessity limit their participation'.[50] A 1962 seventeen-nation conference, expected to set up an Asian version of Euratom, settled for pledging closer ties with the IAEA. Such moves eventually led to a ten-nation 1972 Regional Cooperation Agreement for research, development, and training related to nuclear science and technology, to which Indonesia became party. Under this agreement, the idea of an Asian nuclear research center was revived in 1979, when Japan hosted a five-day meeting at the request of the IAEA on the subject. The result was typically non-committal: no center, but a joint research project on isotope treatment of food.[51] These efforts did not appear to accelerate nuclear power development anywhere; indeed, many Asian interviewees in 1981 pleaded ignorance of the existence of the 1972 agreement.

CONCLUSION

Foreign pressure cannot create nuclear programs. Decisions to initiate nuclear programs in developing countries are made independently. Certainly, deliberations are colored by the example of the nuclear power programs in the advanced nations, but example is distinct from pressure. All developing countries become heavily dependent upon foreign support for what activities they do choose to undertake in the nuclear field. These are the central conclusions of this chapter. Others follow.

First, the success of foreign actors in influencing programs depends upon the customer's receptiveness. Westinghouse could have been much more influential in Iran if the AEOI had pushed for a quick completion of a new agreement for cooperation with the United States. Any potential supplier could have been more influential had the governments in the nonnuclear nations welcomed it. Foreign influence only becomes irresistible once a developing country government has decided the basic issue of whether or not to encourage *any* nuclear power program.

After a commitment is made, developing countries lose flexibility and become more vulnerable to foreign influence. Before a government makes a nuclear commitment, the suppliers need it and lack a lever to pry it to move unwillingly. Once the country is locked into a commitment, foreign pressure can succeed, as when Canada forced Argentina to renegotiate both safeguards and financing terms of the Embalse agreement. Supplier aggressiveness can affect the scale and timing of a program, but cannot force fundamental decisions. If KWU had not so swiftly made such an attractive offer to AEOI, Framatome, the French reactor manufacturer, would have built the first two reactors, if slightly slower and at a higher price. If Framatome stalled, Westinghouse could have filled the breach, and so on. Especially in the buyers' market of today, there is no chance that a single supplier can close down a developing country's nuclear program, or that all would (or even want to) band together to do so. Conversely, nuclear stations are too expensive to be gifts. If a government is unwilling to undertake the financial burden of large capital expenditures (even easy loans must be serviced and repaid, albeit in depreciated currency), then the nuclear suppliers cannot create a viable program.

Second, foreign influence depends upon close collaboration between government and vendor in the supplier country. Two aspects of government support are critical. One is political. Unilateral policies antagonistic to developing countries vitiate a vendor's prospects. The Carter Administration nonproliferation policy ruined Westinghouse and General Electric prospects in Iran, and contributed to their exclusion from the Argentine market. Canadian nonproliferation policy hurt AECL's competitive standing. The other is economic. The expense of these projects requires governments to offer special credit terms if developing nation customers are desired. These concessions are offered by the Export-Import Bank of the United States, the Export Development Corporation of Canada, Hermes Guarantees and the Credit Bank for Reconstruction of West Germany, and the French Company for the Insurance of External Commerce (COFACE). The extent to which governments are willing to deploy the resources of these institutions to support nuclear power exports largely determines its commercial success. Problems with COFACE, for example, contributed to the lengthy delay in the negotiation of the two Ahwaz reactors.

Third, foreign influence is not indiscriminate. Countries with some land, labor, and capital assets conducive to nuclear power are desirable customers. Those lacking these assets are not, and find no suppliers straining to sell them reactors. Modest research support is available. Uranium-poor countries are willing to seek out that mineral, which would greatly assist their own programs were it found in quantity. But limited financing abilities or instability discourages potential suppliers. Besides, other assets may attract investment. In Indonesia, for example, other energy resources encouraged private and public foreign investment in coal (Royal Dutch Shell), natural gas (Mobil Oil), geothermal (New Zealand and the United States), and hydro (Japan) development. In a country limited in skilled personnel and financing capacities, gains for energy alternatives seriously reduce the resources available for nuclear power.

NOTES

1   *Nucleonics Week*, December 26, 1974, p. 4, and January 6, 1977, p. 5. The ban also applied to Euratom and Japan.
2   Mr. Justice W. Fox, *et al., Ranger Uranium Environmental Inquiry: Second Report* (Canberra: Australian Government Publishing Service, 1977), p. 320.
3   See *Presidential Documents – Jimmy Carter*, vol. 13, no. 18, May 2, 1977, and *Nuclear Non-Proliferation Act of 1978*, Public Law 95–242, March 10, 1978.
4   E. S. Milenky, *Argentina's Foreign Policies* (Boulder, Co.: Westview Press, 1978), p. 17.
5   IAEA, *Summary Information on IAEA Safeguards*, unpublished (1976). *Bulletin Today* (Philippines), January 27, 1981.
6   K. Gottstein, 'Nuclear Energy for the Third World,' *Bulletin of the Atomic Scientist*, vol. 33, no. 3 (1977), p. 44.
7   W. Rommel, 'The German Federal Republic in Latin America', *Studies on the Developing Countries*, vol. 2 (Warsaw, 1972), pp. 143–4; and *Nucleonics Week*, December 12, 1968, p. 8.
8   *New York Times*, January 16, 1959.
9   Hon. Clarence D. Long, 'Nuclear proliferation: can Congress act in time?' *International Security*, vol. 2 (Spring 1977), p. 62. See also US Department of State, 'Atomic energy: agreement between the United States of America and Argentina. Effected by exchange of notes, signed at Buenos Aires, November 8, 1962 and November 30, 1963', *Treaties and Other International Acts*, Series 5504 (Washington: US Government Printing Office, 1963).
10 Milenky, *Argentina's Foreign Policies*, op. cit., p. 120.
11 Richard J. Barber Associates, Inc., *LDC Nuclear Power Prospects, 1975–1990: Commercial, Economic and Security Implications*, ERDA-52 (Washington: Energy Research and Development Administration, 1975), p. III-16.
12 Congressional Research Service, Library of Congress, *Nuclear Proliferation Factbook* (Washington: US Government Printing Office, 1980), pp. 301–2; and Long, 'Nuclear proliferation', op. cit., p. 56.
13 R. R. Pelayo, 'Financial aspects of nuclear power programs from the experience of FORATOM member countries', mimeo. (Salzburg: IAEA, 1977), pp. 12–13.
14 M. Kreile, 'West Germany: the dynamics of expansion', *International Organizations*, vol. 31, no. 4 (Autumn 1977), pp. 795–803.
15 Barber, *LDC Prospects*, op. cit., p. IV-36–7.
16 *Nuclear Engineering International*, April 1974, p. 256.
17 J. Surrey and W. Walker, *The European Power Plant Industry*, final revise (Brighton, Sussex: Sussex European Research Centre, May 1981), pp. 32–45.
18 *Times Journal* (Philippines), July 8, 1979, p. 8.
19 Ibid., May 23, 1980.
20 *Nuclear Engineering International*, March 1974, p. 141.

21 *Noticias Argentinas* (Buenos Aires), trans. FBIS, *Worldwide Report: Nuclear*, August 15, 1979.

22 *Nucleonics Week*, September 18, 1975, p. 6; January 22, 1976, p. 5; February 5, 1976, p. 8.

23 KWU Chairman Klaus Barthelt, quoted in *Electrical Review*, June 23, 1978, p. 8.

24 For the Philippines perspective on the NRC consideration and decision to issue the license, see *Philippine Daily Express*, February 2, 1980, and *Bulletin Today* (Philippines), May 8, 1980.

25 *British Nuclear Forum Bulletin* (April 1969), p. 11.

26 UN Document A7677, September 24, 1969, quoted in J. R. Redick, *Military Potential of Latin American Nuclear Energy Programs* (Beverly Hills: Sage Publications, 1972), p. 143.

27 I. Breach, *Windscale Fallout* (Harmondsworth, Middlesex: Penguin, 1978), p. 85.

28 A. Matin, 'Problems in introducing small and medium power reactors to developing countries, with particular reference to the creation of a special nuclear fund', in IAEA, *Small and Medium Power Reactors 1970: Proceedings of a Symposium* (Vienna: IAEA, 1971), p. 495; and Address by IAEA Director General Sigvard Eklund, in AEOI, *Transfer of Technology* (1977 Conference proceedings), (Tehran: AEOI, 1977), vol. 1, p. 26.

29 'Address by President Eisenhower before the UN General Assembly, 8 December 1953', *Congressional Record*, vol. 100, January 7, 1954, pp. 61–3.

30 IAEA, *Statute*, March 1967 reprint, Articles II and III.2; and *IAEA Bulletin*, vol. 15, no. 6 (1973), p. 2.

31 S. Eklund, in AEOI, *Transfer of Technology* (Conference proceedings) (Tehran: AEOI, 1977) vol. 1, pp. 22–6.

32 IAEA, *Small and Medium Power Reactors 1970* (Vienna: IAEA, 1971) and IAEA, *Small and Medium Power Reactors: Conference Proceedings* (Vienna: IAEA, 1961).

33 T. J. Connolly, U. Hansen, W. Jack, and K.-H. Beckurts, *World Nuclear Energy Paths*, for the International Consultative Group on Nuclear Energy (New York and London: The Rockefeller Foundation and the Royal Institute of International Affairs, 1979), p. 40. Three of the fifteen members of the Group were from developing countries.

34 This quotation and most of the information in this paragraph are drawn from the IAEA, *Summary Information Concerning Safeguards* (Vienna: IAEA, 1976).

35 A. Oteiza-Quirno, 'Technical assistance in Latin America', *IAEA Bulletin*, vol. 18, no. 3/4 (1976), p. 25. Concerning UNDP loans to the CNEA, see US Department of State, *Telegram Buenos Aires 949*, February 1979.

36 See technical assistance reports in following numbers of *IAEA Bulletin*: vol. 3 (April 1961), p. 17; vol. 4 (April 1962), pp. 8–9; vol. 5 (April 1963), p. 14; vol. 6 (April 1964), p. 19; vol. 7 (June 1965), p. 28. Also, J. A. Quihillalt, 'El efecto de la cooperación internacional en el plan nuclear argentino', in United Nations and IAEA, *Peaceful Uses of Nuclear Energy: Proceedings of the Fourth International Conference*, vol. 1 (New York: UN and IAEA, 1972), pp. 661–5; and for Euratom agreement, *Nuclear Engineering International* (October 1962), p. 383.

37 *Noticias Argentinas*, trans. FBIS, December 11, 1979.

38 Statement by Dr Sigvard Eklund to the IAEA Board of Governors, *IAEA Press Release* (Vienna), July 6, 1981, p. 4. The ratio of inspectors to facilities, has improved (from 1:8 to 1:4) since 1975, when forty-three inspectors were responsible for 315 facilities, according to the IAEA, *Annual Report for 1975* (Vienna: IAEA, 1976), pp. 190, 44–6.

39 Of the total fiscal year 1980 budget of US$80,643,000, $15,952,000 went to technical assistance, training, and operations, $11,554,000 to research and isotopes, and $19,396,000 to safeguards. IAEA, *The Agency's Account for 1980*

(Vienna: IAEA, July 1981), p. 66, and US Office of Technology Assessment, *Nuclear Proliferation and Safeguards* (Washington: 1977 draft), p. 216.

40 The US government had remonstrated to little avail with both the French and the Italians to restrict their nuclear assistance to Iraq. The Saudis reportedly contributed large sums to the Pakistani enrichment and reprocessing program on condition that Pakistan did not share nuclear weapons technology with the Iraqis.

41 IAEA Press Release, PR 81/10, June 12, 1981.

42 IAEA Press Release, PR 81/9, June 9, 1981.

43 L. Manning Muntzing, 'International instruments for the transfer of nuclear technology', in AEOI, *Transfer of Technology* (1977 Conference proceedings) (Tehran: AEOI, 1977), vol. 3, p. 246.

44 R. Skjöldebrand and R. Krymm, *Nuclear Power: Report to the Government of Iran* (Vienna: IAEA, February 1974), pp. 1–2.

45 S. Eklund, in AEOI, *Transfer of Technology* (1977 Conference proceedings) (Tehran: AEOI, 1977), vol. 1, p. 26.

46 For a postmortem on INFCE, see the International Institute for Strategic Studies, *Strategic Survey 1980–1981* (London: IISS, 1981), pp. 112–13.

47 G. Meyer-Wöbse, 'Nuclear cooperation in the Third World', *Aussenpolitik* (January 1978), pp. 65–74; *The Financial Times* (London), April 6, 1979; and *International Herald Tribune*, March 7, 1977. For President Videla's remarks upon signing the letter of intent, see *The Financial Times* (London), March 8, 1977.

48 *Latin America Weekly Report*, March 27, 1981, p. 3, and A. MacLachlan, 'Argentina: tomorrow's nuclear exporter', *Energy Daily*, September 24, 1980, pp. 3–4.

49 V. S. Karnic, 'Recipient as donor', *India News*, May 14, 1981, p. 10.

50 J. B. Hollister (Director, International Cooperation Administration), 'Using the atom for economic and social progress in Asia', *Department of State Bulletin*, vol. 33, November 7, 1955, pp. 747–50, and Department of State, 'Working group adopts report on Asian regional nuclear center', *Department of State Bulletin*, vol. 37, August 19, 1957, pp. 312–14.

51 *Nucleonics Week*, January 3, 1963, p. 5, and March 21, 1963, p. 5; *IAEA Bulletin*, vol. 17 (October 1975), p. 38; *Kyodo* (Tokyo), October 15, 1979, and October 16, 1979.

11   Conclusion

No easy answers have emerged in the preceding pages. A distressing number of questions have elicited the same frustrating response: it depends. But conditional answers do not a conclusion make. Lurking behind them, general principles which guide and explain particular policies can be discerned. Once recognized, these principles can support judgments on how best to try to cope safely with the spread of civilian nuclear technology.

It is especially difficult to define principles which can explain how a single event can generate opposite effects. Take two examples. The nuclear recession both promoted and discouraged nuclear power in the Third World, forcing suppliers to seek foreign outlets to compensate for the slump in demand, while at the same time fostering doubts within developing countries over whether the nuclear option is attractive. Similarly, the oil crisis simultaneously encouraged resort to nuclear power as an escape from odious dependence upon OPEC, and sapped governments of the financial resources needed to pay for it.

The five principles which follow may help supplier governments design nuclear export policies which are liberal when possible but conservative when necessary. The last two suggest serious problems with how the nuclear export problem has been perceived and confronted. The first three suggest criteria which can identify those governments that can be trusted with liberal access to fission technology. No doubt the notion that governments of the South must earn the trust of those of the North is anathema to the Third World. Like it or not, though, for obvious reasons nuclear trade is intrinsically discretionary and will remain so. Indeed, the effort to ignore or deny that fact has deprived some countries (like India and Pakistan) of important nuclear assistance while giving others (like Switzerland and sometimes West Germany) a pretext for exporting dangerous technologies while piously maintaining that their incontinence is involuntary.

PRINCIPLES

1   Objectives reveal intentions more accurately than do processes
The comparisons in Part Three between security and economic

objectives on one side and domestic and foreign processes on the other suggest that fervent support at home and lavish attention from abroad simply cannot sustain a nuclear power program in a developing country if it is not a central governmental objective there. Nuclear power costs and signifies too much to be controlled by bureaucratic cliques or eager foreign contractors, both of which certainly exert pressure on nuclear policies. Only if it is given pride of place among government projects can nuclear power attain the momentum necessary for success. If it is not, it surely will never materialize by default.

Objectives also explain policies better than do existing conditions in a country. As processes depend upon the political leadership, so can conditions be ignored by it. For example, in trying to correlate interest in nuclear power to national income, one might suppose that there would be greater enthusiasm in Venezuela than in Bangladesh, when in fact the opposite is the case. Efforts to link nuclear activity to available natural or human resources, topography, or other economic indicators prove similarly futile. It is safer to stick to objectives.

This conclusion narrows considerably the range of individuals whose cooperation is essential to a successful nonproliferation policy. It probably is unwise to devote extensive resources to monitoring responsible business practices, beyond the issuance of export licenses and policing of broken export regulations. Though corporations make their interests known, their lobbying is subordinate to primary governmental decisions to export or import nuclear technology. Although Bechtel confided to the Brazilians a willingness to assist with fuel cycle technologies, the US government's veto killed any prospects for such assistance. This conclusion also suggests the natural limits on the power of persuasion; developing country nuclear policies cannot be expected to be more compliant to wishes of the North than are their governments generally. Consequently, liberal export policies are merely accepted as due, without generating the willingness to reciprocate born of gratitude. Perhaps leverage is increased somewhat, but at the expense of showing all other governments that intransigence may not be punished by restricted exports.

2 The relative importance of economic and security objectives depends upon the scale and maturity of the nuclear program

At its early stages, security objectives alone are both necessary and sufficient to justify a nuclear program. They are necessary because any tangible economic benefits are so far down the pike as to be politically irrelevant (if not invisible). Short-term economic considerations weigh heavily against nuclear power, whose large scale and complexity require that a significant fraction of present talent and consumption be diverted in the hope of accelerating production growth. This erodes the appeal of nuclear power for governments still preoccupied with relieving the worst effects of poverty. The 1970s' recession reduced the ability even of less straitened economies to pay for nuclear power, and made nonsense out of many of the optimistic electricity demand projections used to justify it.

On the other hand, security aims are sufficient in the short run to justify a nuclear program, because the important security benefits flow more from research scale activities than from the construction of a nuclear steam supply system. Also, the costs of a nuclear research program can be minimized, while still providing not only some immediate prestige, but also a foothold in a technology of possibly great eventual economic importance.

As it grows, a nuclear program must gather in economic justifications. Security justifications lose force with increased scale, for two reasons. First, marginal security benefits decline as the program becomes commercialized. A research program provides sufficient technology and perhaps fissile material for an explosive. The increase in weapons manufacturing capability provided by a nuclear power station project may not be great. The increase provided by a third or fourth power reactor will probably be negligible. In one way, larger nuclear programs *forfeit* security benefits by increasing the developing country's dependence upon (hence vulnerability to) nuclear suppliers, who use their nuclear leverage to deter weapons development.

Second, the security benefits of a nuclear program are too ambiguous and limited in applicability to justify a $1 billion investment. A nuclear power program, then, must promise significant long-term economic benefits to remain politically viable. Security arrangements are not discarded, but become the guarantor instead of the source of the commitment. (Even while selling a program on economic grounds alone, a government simultaneously buys an option to exploit the military and political security benefits of atomic fission.) Meanwhile, the economic attraction of nuclear power expands along with its scale and time frame. Eventually, oil exporters and importers alike must develop alternative energy sources, and fission is one that offers many advantages. It does not release noisome chemicals and particles into the biosphere. It does release vast quantities of heat which can be converted into electricity more cheaply, according to its proponents, than do other methods. Thus, the two basic justifications for nuclear power are complementary, with security favoring its present and economics its future development.

Ironically, though security and economic *justifications* are mutually reinforcing in promoting a nuclear power program, security and economic *benefits* are mutually conflicting (except where security coincide with economic benefits, as in Taiwan). Governments which seek military or political benefits do so by increasing the ambiguity of their intentions. If they succeed, they fuel suspicions of the nuclear suppliers, who often react by restricting nuclear exports, thereby reducing (if only by delay) the economic benefits provided by nuclear power. Conversely, governments desiring to maximize the economic benefits of nuclear power must bow to increasing nonproliferation conditions imposed by the nuclear suppliers. Each concession reduces the remaining bargaining leverage. Once a government has pledged to safeguard all its nuclear activities by joining the NPT, it can no longer press for access to technology by threatening to build unsafeguarded facilities. It can suggest that it might withdraw from the NPT, but this approach significantly escalates the stakes of the dispute.

This principle reinforces the distinction between the independent and dependent nuclear countries. The former are willing to pay for exploiting security benefits. India sacrificed North American technical assistance and enriched uranium fuel for Tarapur. The Pakistanis lost all outside nuclear assistance, as well as US military and economic assistance for two years. The South Africans lost the supply of highly enriched uranium for the Safari research reactor. The Argentinians were forced to renegotiate safeguards with the Canadians, exacerbating the costly delays in the Embalse project. The Israelis could not take up the American offer of a nuclear power reactor included in the January 1974 Sinai agreement, even if they wanted to do so. The dependents have paid a price, too. By joining the NPT and accepting full-scope safeguards, they have stripped themselves of the ability to exploit ambiguous intentions. Despite Israel's refusal to sign the NPT, in 1980 Egypt ratified the Treaty in order to obtain US power reactors. Even after the Israelis attacked Osirak, the Egyptians proceeded with the initialing of an IAEA safeguards agreement. South Korea agreed not to go through with its reprocessing deal with France, and Taiwan agreed to dismantle a hot-cell laboratory, both at conspicuous American urging. Governments occasionally are willing to swallow their pride to remain in the good graces of the nuclear suppliers. Understanding the tolerance of individual governments in this regard is indispensable to good policy.

3 The type of program selected best illustrates objectives
Words vanish, sometimes without a trace. Capabilities do not, since they are not easily unlearned. Where statements seem inconsistent with capabilities under development, policies should be governed by the latter. Even if a government leader earnestly affirms his peaceful aims, and builds a reprocessing plant solely for waste management purposes, prudence requires that the reprocessing capability should be

treated as dangerous and the statements as possibly disingenuous. Genuinely sincere policies are often reversed, either by their creators or by successive governments. Ultimately the reprocessing plant could come to rest in belligerent hands, unprotected by forgotten assurances. When Saddam Hussein or Zia ul-Haq insist that their aims are peaceful, yet persist in seeking technologies and materials useful to bomb manufacture, engage in secret acquisition of uranium, while actually neglecting to build a nuclear power station, their veracity must be doubted. The dependents' claims are far more credible, as they are backed by large investments. This was even true for the Shah of Iran, who may have harbored nuclear weapon desires, but if so never manipulated his nuclear power program to achieve them. Argentina and India steer a middle course, carefully maintaining a nuclear weapon option and resisting safeguards while continually affirming their peaceful intentions. Like Israel, which has never admitted but is widely credited with nuclear weapons possession, Argentina and India use their ambiguous position to gain status and leverage without incurring the onus of openly military efforts. Deliberate ambiguity does warrant suspicion, but this suspicion should breed isolation, not improved status.

4   Arguments over whether the nonproliferation problem is more technical or political are misguided and destructive

The nonproliferation problem is both technical and political. If each factor is not equally weighty, the difference is insignificant and unquantifiable. Political intentions influence technical capabilities. Japan and West Germany prosper under the *status quo*, and do not wish to disrupt it. That is why, though either government could manufacture nuclear weapons at will, neither is likely to do so in the foreseeable future. They lack nuclear weapons by choice. By contrast, India and Pakistan are poor, yet strong governmental support has given each substantial nuclear explosive programs. Many wealthier or more industrialized nations, in both hemispheres, lack such nuclear capabilities, also from choice.

Conversely, technical abilities influence political intentions. Governments, like individuals, do not attempt where they expect to fail. A strong industrial base or capital surplus encourages the decision to build a nuclear power station. The ability to obtain fissile materials could encourage a government to use them for nuclear weapons. The less effort required, the easier it is to gain approval for a project. A government with a plutonium stockpile faces one less hurdle to the decision to build nuclear weapons. It may still forgo the option, but the nature of its choice has been changed by its capability.

Capabilities are not all internally generated. Developing countries still depend extensively upon the nuclear suppliers for technology, and will continue to do so. The tired refrain, 'nonproliferation is a political, not a technical problem', is simply false, as the INFCE final report confirmed. If technical constraints were not important, then Indonesia probably would have had the bomb in 1965, followed by Libya, Uganda, Iraq, and perhaps others. Technology cannot permanently be dammed behind a set of export controls, but it does not emanate as inevitably or unalterably as a ripple from a pebble dropped in a pond. Technology does not 'spread' of its own accord at all; it must be developed through choice and effort. The rate and direction of its adoption can be influenced, and should be because technological capabilities in turn shape intentions.

5   Inconsistent nuclear supplier policies are harmful

Good bilateral relations improve the nonproliferation regime. Unfortunately, since 1974 inconsistent nuclear supplier policies have marred them. For two decades before Pokharan, nuclear export policies followed the Atoms for Peace premise: atomic assistance to NNWSs can be traded for assurances that assistance will not be used for military purposes. This premise reversed the earlier American policy, expressed in the 1946 McMahon Act, which barred American nuclear assistance even to the Allies. The closed door approach was abandoned because it provoked deep resentment and could not keep the lid on nuclear technology. These problems would return if the Atoms for Peace idea were completely scrapped now, when nuclear knowhow has spread far and wide.

Confusion over Third World intentions after 1974 led nuclear suppliers to adjust their policies. Some change clearly was needed, for the Indians had shown that providing technology would not necessarily deter recipients from nuclear explosive testing, and the proposed exports of reprocessing plants would have facilitated further abuses of peaceful assistance. Canada and Australia restricted uranium exports. All major nuclear suppliers adopted tighter guidelines in the NSG for the application of international safeguards on nuclear activities. France and West Germany pledged not to sell the technology to extract plutonium (usable for weapons) from spent reactor fuel. Controversy persists over whether pre- and post-Pokharan attitudes were too sanguine or too alarmist, respectively. Whichever is the case, it is reasonable to assume that at least the magnitude of the swing in nuclear export policies was unjustified. As a result, suppliers appeared inconsistent and discriminatory, undermining others' faith in the NPT commitment to provide nuclear assistance.

Discrimination is unavoidable. A policy either treats equals differently or unequals the same. Either way, someone is bound to be offended. Good policy requires careful selection of criteria by which to treat governments differently and a politically acceptable justification of those criteria. The Carter Administration's restricted nuclear export policy helped redress the balance toward laxity in nuclear trade, but foundered on a universalism which failed to acknowledge that the same activities in different countries did not equally affect American interests. Enriched uranium supplies to Europe were threatened and even suspended briefly for failure to comply with the NNPA. The Japanese had to agree to limit the use of their new reprocessing facility at Tokai Mura before the US government would permit it. In its quest for fairness, the Carter Administration alienated the allies essential to any successful nonproliferation policy by casting doubt on whether they could rely upon continued American assistance.

Apart from the dangers of policies mismatched to intentions, inconsistency is itself mischievous. First, it destroys credibility. The effectiveness of a policy depends on the belief that its impact cannot be avoided through procrastination. Policy should not be changed frequently or capriciously. By 1977, President Carter was already perceived as inconsistent; many consequently believed that his hard line on nonproliferation (most evident in the Nuclear Non-Proliferation Act of 1978) could be safely ignored until he changed his mind. This cynicism was justified when, two years later, the President overrode a unanimous Nuclear Regulatory Commission recommendation to terminate nuclear cooperation with India under the Act. Second, inconsistency undermines policy by suggesting ulterior motives. For two decades, the United States promoted and dominated the nuclear export markets outside the centrally planned economies. By the early 1970s, however, this market supremacy fell under attack from other nuclear reactor and fuel service suppliers. Even the American monopoly in enriched uranium production was broken, through the French Eurodif and Anglo-Dutch-German Urenco projects. Consequently, when the traditionally energetic American encouragement of nuclear power in the Third World declined, ostensibly because of heightened sensitivity to the link between nuclear energy and nuclear weapons, many reacted distrustfully. American arguments to restrict nuclear exports appeared as a pretext, designed to minimize encroachments on US commercial hegemony in the nuclear market, while giving the United States a breathing space to consolidate its position. Resentful European reactions to the Carter nonproliferation initiatives can be partly understood in this light, while dissent among the nuclear suppliers served the purposes of neither nonproliferation nor alliance solidarity.

Third, inconsistency loses customers. When shopping for a trading partner, the prospective buyer carefully assesses the probability that a contract, once signed, will be fulfilled. Thus, the 1974 American demotion of contracts to supply enriched uranium to Brazil from firm to conditional status damaged the prospects for US Westinghouse (the contractor for the first Brazilian reactor) to sell additional reactors there. In the event, the German firm Kraftwerk Union won the competition to build the next eight Brazilian reactors. Kraftwerk Union's reputation for reliability also influenced the Argentinians, who in 1979 opted for the German proposal to build the Atucha-2 station. Atomic Energy of Canada Limited offered the more thoroughly tested, cheaper CANDU unit, which was more conducive to increased participation of Argentine industry, but Canadian delays and haggling over the construction of a CANDU in Cordoba province had so soured the Argentinians that the German bid prevailed. Sometimes the customer of transferred technology and materials may be lost to a supplier less scrupulous in protecting against misuse.

Fourth, inconsistency facilitates discrimination, an anathema to developing countries. Where exceptions are the rule, rules are useless. Many view the US Nuclear Non-Proliferation Act of 1978 as unduly harsh and abrasive, but it did demonstrate the sincerity of the American commitment to nonproliferation and established a set of guidelines to regularize future nuclear relations. These advantages were gained at the expense of commercial advantage as well as comity in American diplomatic relations, but at least offered the prospect of overcoming the disadvantages of inconsistency. When President Carter excepted India – the only nation which has tested a nuclear device, constructed with the help of foreign peaceful nuclear assistance – this prospect waned. India was assuaged, but a weak signal sent to others, confirming earlier predictions that the United States would not stick to its guns in the matter.

Perfect consistency is unattainable. The effort to reduce inconsistency, however, can at least relieve its worst effects. The unavoidable tradeoff between one inconsistency and another should follow basic principles. The supplier government should draw as clear a line as possible around its vital interests. Are they directly threatened by Japanese and European acquisition of plutonium, by loss of safeguards over Indian nuclear fuel, or only by the detonation of an atomic device? These are thorny but inescapable questions. Once answered, vital interests appear more clearly, and the inconsistencies least obnoxious to them can be adopted.

RECOMMENDATIONS

The principles above help explain the source and significance of nuclear policies in particular developing countries. They also suggest how to confront the dilemma between providing generous access to civil nuclear technology and safeguarding it against abuse for military purposes, at both bilateral and multilateral levels. This section proceeds from the conservative premise that the *status quo* ought to be maintained, and specifically that the acquisition of nuclear weapons by additional states is undesirable. This premise is debatable, but the question of whether a world of more is safer than a world of fewer nuclear powers is beyond the scope of this book.

1   The IAEA should be emphasized but not overloaded

The Agency in many respects already has more extensive powers than any other international organization. Despite its authority to require accountability, to place seals and surveillance equipment on national territory, and to conduct on-site inspections, the IAEA cannot *prevent* the diversion of safeguarded nuclear material to military uses. Its goal is the 'timely detection' of safeguards abuses, but the Agency can do no more than report violations to its Board of Governors and the UN Security Council. Timely detection is only helpful if it enables someone to prevent or deter the potential proliferator. It has not been successful in Pakistan, where four years after its surreptitious nuclear program came to light, nothing has been able to arrest the project.

Since the Agency can neither prevent diversions nor punish transgressors, some have advocated that its powers be strengthened. To be sure, safeguards can and should be technically improved. Significant extension of the IAEA mandate, however, is unlikely. Governments yield sovereignty reluctantly, and the Agency already has offended many developing countries by its intrusiveness, making further cessions of sovereignty politically difficult. Proposals to eliminate the conflict of interests inherent in an agency which both regulates and promotes, by dividing the IAEA in two, have equally poor prospects. Unlike the US AEC, which in 1974 was divided into the Nuclear Regulatory Commission and the promotional Energy Research and Development Administration, the IAEA is the product of a bargain, not of the sovereign disposition of constitutional authority. Carrot cannot be divided from stick without possibly fatal effects for each. Recipients will not accept safeguards without incentives. Suppliers will not provide incentives without safeguards.

The IAEA, then, must be accepted more or less as is. Despite its defects, the Agency embodies an invaluable consensus that nuclear energy should be made available only for peaceful uses. That consensus has lasted for a quarter of a century and remains the foundation of worldwide efforts to arrest the spread of nuclear weapons. The system relies first and foremost upon good faith, which must not be compromised.

2   The Non-Proliferation Treaty should be strengthened

The NPT resembles the IAEA in many ways. It cannot prevent nuclear weapons proliferation. Governments can renounce it or evade its obligations. Despite these constraints, it is seminally important to efforts to halt the spread of nuclear weapons, because it has entrenched a worldwide presumption that nuclear weapon proliferation is bad. This presumption has increased the political cost for *any* government – even if it refuses to sign the Treaty – to go nuclear.

Also like the IAEA, the NPT draws its legitimacy from its breadth of support. The Agency has 110 members, and the Treaty has 115 parties. Unfortunately, that legitimacy is not universal and is now threatened. Important NPT holdouts remain, including two of the five official nuclear weapon states (China and France), and most of the worrisome threshold nuclear powers (Argentina, Brazil, India, Israel, Pakistan, and South Africa). The nuclear weapon states party to the Treaty – Great Britain, the Soviet Union, and the United States – have weakened it by their poor performance in fulfilling obligations under Article IV to facilitate peaceful nuclear technology exchange and under Article VI to negotiate effective arms control measures. Lack of progress on Article VI more than any other factor led to the failure in 1980 at the second NPT Review Conference to agree on any concluding statement. Lack of superpower arms control will not directly lead nonnuclear weapon states to acquire nuclear weapons. That decision is made by assessing overall security interests, not by reaction to increases in arsenals which may have little if any effect on the security of a Third World government. Nevertheless, disregard of Article VI betrays a contempt for NPT obligations which could prove contagious.

The NPT could also be gravely threatened by failure to implement Article IV obligations, which require the NWSs to facilitate the fullest possible exchange of nuclear technology. Article IV was a major attraction to NPT membership, for it pledged that governments which forswore the nuclear option would receive preferential treatment in nuclear technology transfers. This was a more tangible incentive than that offered in Article VI. The problem is that, in practice, NPT parties do not receive preferential treatment. If anything, they are taken for granted, while all sorts of special offers and exceptions are made for those nations which most adamantly refuse to commit themselves to nonproliferation agreements and most assiduously avoid international safeguards. Countries on the threshold of nuclear weapons status reap the extra benefits of using their non-NPT status repeatedly to obtain concessions from nuclear suppliers, whereas NPT parties are implicitly penalized for their allegiance. Argentina and India, though they have paid in some ways for their independence, have never been deprived of the nuclear technology they desired. Why should they accede to the NPT, only to lose the ability to pry more concessions from the nuclear suppliers and to condemn the discriminatory world order?

Conversely, what advantages does NPT adherence confer? Article V, which promised assistance for PNEs, has become a dead letter. The arms control measures promised in Article VI, if they ever materialize, would benefit NPT parties no more than nonparties. And Article IV also entails no privileges to parties over nonparties, while Articles I through III strip NNWSs of bargaining leverage over nuclear weapon acquisition and safeguards acceptance. Unless palpable advantages to NPT membership appear, the prospect for new adherents will remain dim, while the prospect for renunciation of the Treaty increases. Widescale defections from the NPT do not appear imminent, but experiences at the Second NPT Review Conference should chasten those who believe that defections are impossible.

The final serious threat to the nonproliferation regime is posed by the possibility of destabilizing events, such as the acquisition or detonation of nuclear devices by a NNWS or a nongovernmental group. If new nuclear powers are seen to gain status with impunity, or at acceptable political cost, future efforts to deter proliferation will be jeopardized. Events may be ambiguous. The mysterious flashes spotted in the South Atlantic in September 1979, identified by some as nuclear weapons tests, present a dilemma. If you deny their importance, the nonproliferation regime appears to be doing nothing to stop nuclear weapons deployment by certain nations, widely suspected to be Israel or South Africa. On the other hand, if you admit their importance, it is necessary to decide what sanctions should be applied and to whom, when responsibility is denied and the evidence is at best inconclusive. The unhappy history of sanctions suggests that the attempt to apply them may merely highlight the ineffectuality of the system.

Consequently, the NPT cannot simply be preserved as is. If it is not strengthened, it may begin to be renounced or ignored. The proper policy prescription is as obvious as it is difficult to fill: the great powers should harness their nuclear arsenals, and NPT membership should confer material advantages to nonnuclear weapon states. The NNWS NPT-parties should be given preferential access to nuclear technology, equipment, and materials. They should also be rewarded by military, political, and economic support for their legitimate security interests. (Of course, nuclear policy is not the only issue of diplomatic concern, and these rewards must reflect the overall relations among the governments concerned.) The nonnuclear countries cannot be expected to remain that way if cooperation costs rather than confers bargaining leverage. The NPT must benefit its adherents if governments are to be prevented from building unsafeguarded facilities, like the ones now used by the independents to increase their international influence. Some guidelines for these nuclear exports are suggested in the final recommendation.

3 Preferential treatment to NPT holdouts should cease

This is a necessary corollary to the previous recommendation. Always conceding special status and NNPA exemptions to India could gravely undermine the NPT. Such a perverse system of appeasing NPT holdouts conveys the wrong impression to nonnuclear countries: that they can make the greatest impact and obtain the most solicitous attention by flouting the NPT. After all, the worrisome NPT holdouts are those most able to obtain nuclear weapons. The phase of preventing them from attaining weapons capability has passed. So long as India can produce plutonium at Trombay, improving safeguards at its power stations has only marginal importance. On the other hand, all of the holdouts still depend upon nuclear suppliers in one way or another, and their vulnerability should be exploited to prevent the execution of nuclear threats.

This is not to say that we should give up on cordial approaches to the NPT holdouts or other threatening states. As suggested below, alternative arrangements should be employed wherever possible. Where none succeed, the game still is not lost. Nuclear weapons acquisition, as distinct from nuclear threats, does not depend mainly upon nuclear supplier policies for Tarapur or Safari fuel, or Argentine heavy water, but rather on the governments' overall security calculations, which are unlikely to be decisively swayed by Western or Japanese civil nuclear export policies. The NPT and IAEA regime should attend more to its members, many of whom soon will have the same capabilities that the worrisome holdouts have now. It is wiser to bind a government tightly into nonproliferation commitments while it remains more dependent and tractable than it may become. If confidence in the system can be restored this way, the appeal of joining the ranks of the outsiders will fade.

4 Other nonproliferation agreements should be promoted

This recommendation flows from the last two. The NPT should be strengthened, but it can never become a talisman, for two reasons. First, even apart from the inadequate NWS performance under Articles IV, V, and VI, the NPT can plausibly be condemned as discriminatory. This argument is made by IAEA members who have refused to accede to the NPT. They contend that their IAEA membership confers full privileges for access to peaceful nuclear technology, and that the NPT cannot be used to deny them this privilege. The dilemma in any debate pitting Agency against Treaty is that one or both may suffer. It may be wiser to sidestep the legalities of this issue and simply treat the NPT as the natural progeny of the IAEA, introduced at a time when the dangers of indigenously built, unsafeguarded facilities had grown much worse than they had been in 1957. NPT safeguards were consequently strengthened over the earlier IAEA version, and it is logical that the preferences entailed in NPT membership should be commensurately increased.

Second, NPT effectiveness is inherently limited. The Treaty contains some glaring loopholes. It cannot, for example, prohibit nonparties from transferring technology to NNWS parties to the Treaty. It does not bar nuclear weapon *testing* at all. Other agreements, such as no-use pledges or nuclear weapon-free zones can plug these loopholes. More obviously, the NPT is completely inoperative against holdouts, some of whom will always be with us. That is why some analysts urge that holdouts be mollified, not punished, in order to keep them in check. They fear that if the suppliers do not offer a holdout a desired facility under safeguards, then it may build the facility itself. It may be of poor quality and take longer to build, but that indigenous facility will be more dangerous because unsafeguarded.

These NPT deficiencies argue powerfully for resort to alternative methods of control. Safeguards should be applied wherever possible. Argentina should be pressed to fulfill its promise to ratify the Treaty of Tlatelolco. Other NWFZs and a comprehensive test ban should be promulgated. Security assurances to Israel and Pakistan should be linked to abstinence from nuclear weapons testing. The more varied the available menu of nonproliferation commitments, the greater is the likelihood that more governments will find an acceptable one.

The main drawback to this diversified approach is that every attractive alternative undercuts the comparative appeal of NPT membership. Since universal NPT adherence is unattainable, this problem to some extent must be endured if NPT holdouts are to be at all constrained. This endurance should expire at the point where attention to the wishes of the intransigent governments subordinates concerns for the rest.

5   *Clear export criteria should be adopted*

This recommendation addresses the harmful inconsistencies noted above. Two critical criteria are: (1) governmental nonproliferation credentials, and (2) the intrinsic sensitivity of the materials, equipment, or technology proposed for export. Neither category permits black and white distinctions. Several increments exist between NPT membership and an open bid for the bomb, as they do between low enriched uranium and pure plutonium. Between these poles, ideally one would establish clear guidelines of increasingly trustworthy credentials and increasingly sensitive technologies. The better a government's credentials, the better access to technology it should be allowed.

At the extremes of both axes, bright lines will benefit policy. Governments that have accepted a comprehensive nonproliferation constraint – such as the NPT or a no-use pledge – should equally be given most favored nation status in the nuclear field. Those that reject all constraints should receive no nuclear technology. Technologies and materials that are inherently dangerous, such as plutonium reprocessing or 90 percent enriched uranium, should not be exported at all. Completely innocuous technologies and materials should be available to any government that accepts any major nonproliferation commitment, such as IAEA membership. Between these political and technological extremes, credentials possessed should be weighed against technology desired, in deciding whether to approve an export. The consultation mechanism of the 1978 NSG guidelines and perhaps the INFCE-spawned Committee for Assurance of Supply should be used to maximize consistency within these discretionary bounds.

Where they are drawn, lines should be clear and decisive. A government with most favored nuclear trade status should be permitted to import any nuclear technology, except those universally forbidden. Its orders should not be detained while, say, the US NRC reassesses the entire situation for each and every license request. Expeditious licensing should be used to encourage accession to nonproliferation commitments.

The use of these export criteria would undoubtedly be discriminatory, but sensibly so. The differing attitudes of developing countries, though they overlap, at least offer a broad, supportable standard for designing a policy which rewards some and punishes others. One can reasonably treat governments differently, for instance, on the basis of their willingness to submit to nonproliferation accords and to refrain from activities which needlessly aggravate international apprehensions, such as the construction of unsafeguarded plutonium reprocessing plants. In practice, only the independent governments would be likely to be treated harshly under these criteria. The nonnuclear governments would largely be unaffected. Governments truly interested in only the civilian uses of nuclear technology probably will not let the proscription of a few items, which are far more useful militarily than civilly, undermine their entire program. The Shah, for example, eventually yielded to the Americans on the reprocessing issue rather than risk losing the partnership.

It is dangerous to *promote* nuclear power in order to restrain nuclear weapons proliferation. Without nuclear power, the arms controller's task would be much simplified. But since desires for nuclear power in many quarters cannot be wished away, the nuclear energy option should be accepted. So long as a government adheres to the NPT or equivalent in letter and spirit, nuclear commerce should not be hindered. To suppress nuclear technology revives its mystique, arouses envy and pride, and so enhances the appeal of the forbidden fruit. Left alone, its cost suffices to deter most governments from capricious entry into a nuclear power program.

The future of nuclear power will influence the prospects for proliferation. On the one hand, the spread of civil nuclear technology unavoidably spreads nuclear weapons capabilities. On the other, when nuclear power thrives, neither suppliers nor recipients wish to disrupt nuclear trade, and so there is an added incentive for buyers to refrain from pursuing identifiably military nuclear programs. In a sense, then, it is better for nonproliferation objectives if the world nuclear industry is prospering, and eager to avoid disruptions, or dead, so that the spread of nuclear technology would have unambiguously sinister connotations. Most dangerous is the present, limping nuclear industry, where technology becomes more widely dispersed without a commensurate incentive for responsible behavior.

Meanwhile, the Atoms for Peace concept should be widened to include other forms of energy. Title V of the Non-Proliferation Act sought to increase American assistance to meet developing country energy needs, reduce their dependence upon petroleum fuels, and expand their available energy alternatives. International organizations and environmentalists also lobby for nonnuclear energy alternatives. These efforts have been well received in some developing countries, and should be continued because they help Third World governments face their energy problems with more information about more alternatives. Better information facilitates better policy.

These recommendations emerge from the central conclusion of this book, that different nuclear power policies express different governmental goals. Objectives tell more about the reasons behind nuclear power policies than do the processes of domestic politics or foreign influence, which reflect more than shape these goals. Security aims are paramount to governments. They can provide a foundation for support, but cannot carry a program far in the absence of the physical, human, and financial resources necessary to make it economical. Countries do not glide into a nuclear program. With the blessing of the executive, an atomic energy commissioner can overcome strong domestic political opposition. Without it, the cause is hopeless. Nor can foreigners push a policy far in directions inconsistent with leadership objectives. Despite the vested interests often clustered round them, nuclear power policies evolve not from bureaucratic accident, inducement, or compulsion, but from free choice.

Selected Bibliography

PRIMARY SOURCES

I International Documents

Aráoz, A. and Martinez Vidal, C., *Ciencis e industria: un caso argentino* (Washington, D.C.: General Secretariat of the Organization of American States, 1974).

'Argentina', *IAEA Bulletin*, vol. 14, no. 6 (1972).

International Atomic Energy Agency, *The Agency's Account for 1980* (Vienna: IAEA, July 1981).

International Atomic Energy Agency, *The Annual Report for 1975* (Vienna: IAEA, July 1976).

International Atomic Energy Agency, *Market Survey for Nuclear Power in Developing Countries: General Report* (Vienna: IAEA, 1973).

International Atomic Energy Agency, *Market Survey for Nuclear Power in Developing Countries: 1974 Edition* (Vienna: IAEA, 1974).

International Atomic Energy Agency, *Nuclear Power Planning Study for Indonesia* (Vienna: IAEA, 1976).

International Atomic Energy Agency, *Problems Associated with the Export of Nuclear Power Plants: Proceedings of a Symposium* (Vienna: IAEA, 1978).

International Atomic Energy Agency, *Small and Medium Power Reactors: Conference Proceedings* (Vienna: IAEA, 1978).

International Atomic Energy Agency, *Small and Medium Power Reactors 1970: Proceedings of a Symposium* (Vienna: IAEA, 1971).

International Atomic Energy Agency, *Statute* (N.p: March 1967 reprint).

International Nuclear Fuel Cycle Evaluation, *Final Report* (Vienna: IAEA, 1980).

International Nuclear Fuel Cycle Evaluation, *Fuel and Heavy Water Availability: Report of Working Group 1* (Vienna: IAEA, 1980).

Sabato, J., Wortman, O. and Gargiulo, G., *Energía atómica e industria nacional* (Washington, D.C.: General Secretariat of the Organization of American States, 1978).

United Nations, Department of International Economic and Social Affairs. Statistical Office. *World Energy Supplies* (New York: United Nations, 1979).

United Nations, Industrial Development Organization, *Guidelines for Project Evaluation* (New York: United Nations, 1972).

US Arms Control and Disarmament Agency, 'Treaty for the prohibition of nuclear weapons in Latin America, February 14, 1967', in *Arms Control and Disarmament Agreements* (1977 edn) (Washington, D.C.: US Arms Control and Disarmament Agency, 1977).

US Arms Control and Disarmament Agency, 'Treaty on the Non-Proliferation of Nuclear Weapons, July 1, 1968', in *Arms Control and Disarmament Agreements* (1977 edn) (Washington, D.C.: US Arms Control and Disarmament Agency, 1977).

World Bank, *World Development Report 1980* (New York: Oxford University Press, 1980).

II National Documents and Other Official Statements

A Argentina

Alonso, M., 'Nuclear developments in Latin America and the fuel cycle', paper presented at the Conference on International Commerce and Safeguards for Civil Nuclear Power, New York, March 13–16, 1977. Sponsored by the Atomic Industrial Forum.

Allegria, J. L., Coll, J. A. and Quihillalt, O. A., 'El efecto de la cooperación internacional en el plan nuclear argentino', in United Nations and IAEA, *Peaceful Uses of Atomic Energy: Proceedings of the Fourth International Conference*, Vol. 1 (New York: United Nations, 1972), pp. 661–5.

Allegria, J. L., Coll, J. A. and Quihillalt, O. A., 'La contribución de la energía nuclear a la solución del problema energético argentino', in United Nations, *Peaceful Uses of Atomic Energy: Proceedings of the Third International Conference*, Vol. 1 (New York: United Nations, 1965), pp. 104–10.

Allegria, J. L., Coll, J. A. and Quihillalt, O. A., *Una breva resena histórica de la CNEA* (Buenos Aires: Sociedad Científica Argentina, n.d.).

Aráoz, A., 'Actividades de desarollo para la fabricación de elementos combustibles en la Argentina', in *United Nations, Peaceful Uses of Atomic Energy: Proceedings of the Fourth International Conference*, Vol. 8 (New York: United Nations, 1972), pp. 245–61.

Baez, J. N. *et al.*, *Participación de la industria argentina en la central nuclear en Atucha y futuras* (Buenos Aires: CNEA, 1973).

Bronstein, B., 'Argentina's hydroelectric-based energy plan', *Power Engineering*, January 1981, pp. 50–1.

Cosentino, J. O., 'The Argentine nucleoelectric programme', mimeo., unpublished (1979).

Csik, B. J., 'Posibilidades de centrales nucleares de pequeña o mediano potencia en Argentina', in International Atomic Energy Agency, *Small and Medium Power Reactors, 1970. Proceedings of a Symposium* (Vienna: IAEA, 1971), pp. 431–8.

Frydman, R. J., Wortman, O. and Sabato, J. A., 'Towards an Argentine nuclear industry', in Atomic Energy Organization of Iran, *Transfer of Technology* (1977 Conference proceedings) (Tehran: AEOI, 1977), Vol. 2, pp. 185–91.

Kaufmann, F., *Planning a Fuel Reprocessing Plant* (Buenos Aires: CNEA, 1973).

Quihillalt, O. A., 'Argentine experiences in its nuclear power programme', mimeo., unpublished (1979).

República Argentina, Comision Nacional de Energía Atómica, *Actividades de la Gerencia de Tecnologia* (Buenos Aires: CNEA, August 1973).

República Argentina, Comision Nacional de Energía Atómica, *Feasibility Study: Nuclear Power Plant for the Greater Buenos Aires-Littoral Area: Summary*, unpublished (1965).

República Argentina, Comision Nacional de Energía Atomica, *Memoria [Anual]*, Vols 1963–76 (Buenos Aires: CNEA, 1963–76).

República Argentina, El Poder Ejecutivo Nacional, *Decreto No. 10.936*, May 31, 1950.

República Argentina, El Poder Ejecutivo Nacional, *Decreto No. 384*, October 6, 1955.

República Argentina, El Poder Ejecutivo Nacional, *Decreto Ley No. 22.477*, December 18, 1956.

República Argentina, El Poder Ejecutivo Nacional, *Decreto Ley No. 22.498*, December 1956.

República Argentina, El Poder Ejecutivo Nacional, *Decreto No. 5.423*, May 23, 1957.

República Argentina, El Poder Ejecutivo Nacional, *Decreto No. 842*, January 24, 1958.

República Argentina, El Poder Ejecutivo Nacional, *Decreto No. 7006*, June 10, 1960.

República Argentina, El Poder Ejecutivo Nacional, *Decreto No. 485*, January 22, 1965.

República Argentina, El Poder Ejecutivo Nacional, *Decreto No. 749*, February 20, 1968.

República Argentina, El Poder Ejecutivo Nacional, *Decreto No. 7619*, December 4, 1968.

República Argentina, El Poder Ejecutivo Nacional, *Decreto No. 3183*, October 19, 1977.

República Argentina, El Poder Ejecutivo Nacional, *Decreto No. 302*, January 29, 1979.

República Argentina, El Poder Ejecutivo Nacional, *Decreto No. 2441*, September 28, 1979.

República Argentina, Ministerio de Economia, Secretaria de Estado de Energia, *Plan nacional de equipamiento para los sistemas de generación y transmision de energia electrica periodo 1979–2000*, 4 Vols (Buenos Aires: 1979).

Ruda, J. M., 'La posición argentino en cuanto a Tratado sobre la No Proliferación de las Armas Nucleares', *Estrategia* (September 1970–February 1971), p. 79.

Sabato, J. A., 'Atomic energy in Argentina: a case history', *World Development*, vol. 1 (August 1973), pp. 23–37.

Sabato, J. A., 'Energía atómica en Argentina', *Estudios Internacionales* (Santiago), vol. 20 (October–December 1968), pp. 332–57.

Sabato, J. A., Interviewed by Robert Lindley, *Washington Star*, October 6, 1975.

Sabato, J. A. and Frydman, R. J., 'La energía nuclear en America Latina', *Estrategia* (September–October 1976), pp. 54–62.

United States, Department of State, *Airgram Buenos Aires A-128*, August 30, 1976.

United States, Department of State, *Airgram Buenos Aires A-15*, February 9, 1977.

United States, Department of State, *Airgram Buenos Aires A-54*, May 10, 1977.

United States, Department of State, *Telegram Buenos Aires 00304*, January 16, 1976.

United States, Department of State, *Telegram Buenos Aires 01462*, March 4, 1976.

United States, Department of State, *Telegram Buenos Aires 03492*, May 26, 1976.

United States, Department of State, *Telegram Buenos Aires 04079*, June 21, 1976.

United States, Department of State, *Telegram Buenos Aires 7027*, August 28, 1979.

United States, Department of State, 'Atomic energy: cooperation for civil uses: agreement between the United States of America and Argentina, signed at Washington, June 22, 1962, with exchange of notes', *Treaties and Other International Acts*, Series 5125 (Washington, D.C.: US Government Printing Office, 1963).

United States, Department of State, 'Atomic energy: equipment for use at La Plata University: agreement between the United States of America and Argentina, effected by exchange of notes, signed at Buenos Aires, November 8, 1962 and November 30, 1963', *Treaties and Other International Acts*, Series 5504 (Washington, D.C.: US Government Printing Office, 1963).

B   Iran

Atomic Energy Organization of Iran, *Proceedings of the Conference on Transfer of Nuclear Technology, Persepolis, Shiraz, Iran*, 4 Vols (Tehran: AEOI, 1977). See articles by Arabian, Moshfegh-Hamadani, Nooshin and Mehraban, Schmidt, Sioshansi, Sohrabpour, Sotoodehnia, and Vakilian.

Atomic Energy Organization of Iran, Nuclear Research Center, *Progress Report, October–December 1975* (Tehran: AEOI, January 1976).

Etemad, A., interview in *Nucleonics Week*, July 15, 1976, pp. 7–8.

Etemad, A. and Manzoor, C., 'Le programme electronucléaire de l'Iran', *Annales des Mines* (May–June 1978).

Mossavar-Rahmani, B., 'Iran's nuclear power programme revisited', *Energy policy*, vol. 8, no. 3 (September 1980).

Pahlavi, His Imperial Majesty Reza Shah, Shahanshah of Iran, *Mission for My Country* (London: Hutchinson, 1961).

Pahlavi, His Imperial Majesty Reza Shah, Shahanshah of Iran, interviewed in R. K. Karanjia, *The Mind of a Monarch* (London: George Allen & Unwin, 1977).

Sarrami, M., Sadri, N. F. and Etemad, M. A., 'Economic promise – nuclear power in Iran', *Transactions: American Nuclear Society*, vol. 24 (1976), pp. 364–5.

Skoldebrand, R., and Krymm, R., *Nuclear Power: Report to the Government of Iran* (Vienna: IAEA, February 1974).

United States, Department of Energy, 'Atomic energy program of Iran', mimeo. (Washington: n.p., 1978).

United States, Energy Research and Development Administration, 'Iran: atomic energy program', mimeo. (Washington: n.p., October 1976).

United States, Department of State, 'Atoms for Peace agreement with Iran', *Department of State Bulletin*, vol. 36 (April 1957), pp. 629–30.

United States, Department of State, *Airgram Tehran A-14*, January 26, 1976.

United States, Department of State, *Telegram Tehran 00385*, January 15, 1976.

United States, Department of State, *Telegram Tehran 01232*, February 7, 1977.

C   *Indonesia*

Akil, I., 'Development of geothermal resources in Indonesia', *Proceedings of the Second UN Symposium on Development and the Use of Geothermal Resources* (Berkeley: University of California, 1976).

Arismunandar, A., *The Indonesian Energy Demand in the Year 2000: Views and Comments* (Jakarta: PLN, March 1975).

Arismunandar, A., Speech reported in *Kompas* (Jakarta), translated by Foreign Broadcast Information Service, October 18, 1979, p. 13.

Baiquni, A., interviewed in *Kompas*, translated by Foreign Broadcast Information Service, June 1, 1979, p. 12.

Baiquni, A. and Sudarsono, B., 'First steps toward a nuclear power programme: the Indonesian experience and prospects', *Proceedings: The First Pacific Basin Conference on Nuclear Power Development and the Fuel Cycle* (Honolulu: American Nuclear Society, October 11–14, 1976), pp. 43–52.

BPP Teknologi, with Bechtel National, Inc., *Alternative Strategies for Energy Supply in Indonesia 1979–2003* (Jakarta: n.p., December 1980).

Iljas, J., 'Pembentukan tenaga ahli dan tehnik untuk pusat listrik tenaga nuklir pertama', *Majalah BATAN*, vol. 7 (1974), pp. 2–13.

Iljas, J. and Subki, I., 'Nuclear power prospects in an oil and coal-producing country', in IAEA, *Nuclear Power and Its Fuel Cycle: Proceedings of an International Conference*, Vol. 6 (Vienna: IAEA, 1977), pp. 101–13.

Italo-Indonesian Workshop on Industrial Participation in the Implementation of a Nuclear Power Program, 'Feasibility study for the first nuclear power plant in Indonesia', mimeo. (Jakarta: NIRA-ENEL and BATAN-PLN, March 1979).

Republic of Indonesia, Badan Tenaga Atom Nasional, *Master Plan* (Jakarta: BATAN, May 1972).

Republic of Indonesia, Badan Tenaga Atom Nasional, *Seminar teknologi dan ekonomi pusat listrik tenaga nuklir* (Bandung: BATAN, 1971).

Republic of Indonesia, Badan Tenaga Atom Nasional, *Teknologi pusat listrik tenaga nuklir* (Jakarta: BATAN, 1970).

Republic of Indonesia, Embassy of the Republic of Indonesia, Washington, D.C., *Indonesia Develops: Five-Year Development Plan, April 1969-April 1974*, 1970 reprint (Washington: n.d.).

Republic of Indonesia, Embassy of the Republic of Indonesia, Washington, D.C., *Second Five-Year Development Plan (April 1974-March 1979): A Brief Description* (Washington: n.d.).

Republic of Indonesia, Embassy of the Republic of Indonesia, Washington, D.C., *REPELITA III: The Third Five-Year Development Plan* (Washington: n.d.).

Republic of Indonesia, Institute of Atomic Energy, *Atoms for Peace in Indonesia* (1963).

Shepherd, J. H., *Prospecting for Radioactive Minerals in Indonesia: Report to the Government of Indonesia* (Vienna: IAEA, 1963).

Siwabessy, G. A., 'Statemen direktur djenderal BATAN', *Majalah BATAN*, vol. 3, no. 4 (1970), pp. 3–7.

Sudarsono, B., 'The need for nuclear energy in Indonesia', *Indonesian Quarterly*, vol. 8, no. 3 (July 1980), pp. 62–70.

Sudarsono, B. and Prayoto, Ir., 'The role of nuclear power in Indonesia's power planning', in American Nuclear Society, *Transactions: The Second Pacific Basin Conference on Nuclear Power Plant Construction, Operation, and Development*, vol. 29 (1978), pp. 10–15.

Suharto, 'Sambutan pada ulang tahun ke-X BATAN dan peresmian penggunaan cobalt irradiator', *Majalah BATAN*, vol. 2, no. 1 (1969), pp. 1–2.

Supadi, S., Subki, I. and Surjadi, A. J., 'Status report of the Bandung Reactor Center', in IAEA, *Research Reactor Utilization* (Vienna: IAEA, 1972), pp. 33–7. See also article by Subki and Linggoatmodjo, pp. 275–7.

United States, Department of State, *Airgram Jakarta A-61*, April 23, 1975.

United States, Department of State, *Airgram Jakarta A-59*, June 13, 1978.

United States, Department of State, *Telegram Jakarta 00810*, January 20, 1976.

United States, Department of State, 'Using the atom for economic and social progress in Asia', by John Hollister, *Department of State Bulletin*, vol. 33, November 7, 1955, pp. 747–50.

United States, Department of State, 'Working group adopts report on Asian regional nuclear center', *Department of State Bulletin*, vol. 37, August 19, 1957, pp. 308–14.

United States, Energy Research and Development Administration, 'Indonesian atomic energy program', mimeo (September 1975).

D   Other Asia

Atomic Energy Council, Executive Yuan [Taiwan], *The Development of the Atomic Energy Programme of the Republic of China* (Taipei: n.d.).

Ministry of Economic Affairs, Energy Committee, *Energy Policy for the Taiwan Area, Republic of China*, approved by the Executive Yuan, January 11, 1979.

Ministry of Economic Affairs, Energy Committee, *The Energy Situation in Taiwan, Republic of China* (Taipei: n.p., March 1980).

National Power Corporation [Philippines], *1979 Annual Report* (Manila: n.d.).

National Power Corporation [Philippines], *Philippine Nuclear Power Plant* (Manila: NPC, 1978).

Republic of the Philippines, Ministry of Energy, *Ten-Year Energy Programme 1980–1989* (Manila: Planning Service, Ministry of Energy, 1980).

Sethna, H. N., 'India's atomic energy program – past and future', *IAEA Bulletin*, vol. 21, no. 5 (October 1979), pp. 2–11.

Taiwan Power Company, *Taipower and Its Development* (Taipei: n.p., 1980).

Taiwan Power Company, *Taipower '79* (Taipei: n.d.).

E   United States

Barber Associates, R. J., Inc., *LDC Nuclear Power Prospects, 1975–1990: Commercial, Economic and Security Implications*, ERDA-52, prepared for the US Energy Research and Development Administration under Contract No. AT(49-1)-3665 (Washington, D.C.: US ERDA, 1975).

Fox, J. B., *et al., International Data Collection and Analysis*, 6 Vols, prepared for the US Department of Energy under Contract No. EN-77-C-01-5072 (Atlanta: Nuclear Assurance Corporation, April 1979).

United States, *Nuclear Non-Proliferation Act of 1978*, Public Law 95-242, March 10, 1978.

United States, Department of Energy, Office of International Affairs, *International Petroleum Annual* (Washington, D.C.: US Government Printing Office, 1980).

United States, Department of Energy, Office of International Affairs, *The Role of Foreign Governments in the Energy Industries* (Washington, D.C.: US Government Printing Office, October 1977).

United States, Library of Congress, Congressional Research Service, *Nuclear Proliferation Factbook*, 1977 and 1980 edns (Washington, D.C.: US Government Printing Office, 1977 and 1980).

United States, Office of Technology Assessment, *Nuclear Proliferation and Safeguards* (Washington: n.p., 1977 draft).

United States, *Presidential Documents: Jimmy Carter 1977*, Vol. 13, April 18, 1977.

SECONDARY SOURCES

I   Books, Articles, and Reports

A   General, Energy, and Economic

Allison, G., *The Essence of Decision* (Boston: Little-Brown, 1971).

Anderer, J., McDonald, A. and Nakicenovic, N., *Energy in a Finite World: Paths to a Sustainable Future* (Cambridge, Mass.: Ballinger, 1981).

Babha, H. J., 'Science and the problems of development', *Science*, vol. 151 (February 4, 1966), pp. 541–9.

Calabresi, G. and Bobbitt, P., *Tragic Choices* (New York: W. W. Norton, 1978).

Collingridge, D., *The Social Control of Technology* (London: Frances Pinter, 1980).

Eckholm, E., *Losing Ground* (New York: W. W. Norton, 1976).

Hirschman, A. O., *Development Projects Observed* (Washington, D.C.: Brookings Institution, 1967).

Kreile, M., 'West Germany: the dynamics of expansion', *International Relations*, vol. 31, no. 4 (Autumn 1977), pp. 795–803.

Landsberg, H. H., *et al., Energy: The Next Twenty Years* (Cambridge, Mass.: Ballinger, 1979).

Little, I. M. D. and Mirrlees, J. A., *Project Appraisal and Planning for Developing Countries* (London: Heinemann Educational Books, 1974).

Lovins, A. B., *Soft Energy Paths: Toward a Durable Peace* (Cambridge, Mass.: Ballinger, 1977).

Robinson, A. (ed.), *Appropriate Technologies for Third World Development* (New York: St. Martin's, 1979).

Smil, V. and Knowland, W. E. (eds), *Energy in the Developing World: The Real Energy Crisis* (Oxford: Oxford University Press, 1980).

Sorensen, T. C., *Kennedy* (London: Hodder & Stoughton, 1965).

Stewart, F., *Technology and Underdevelopment* (London: Macmillan, 1977).

Surrey, J. and Walker, W., *The European Power Plant Industry*, final revise (Brighton, Sussex: Sussex European Research Centre, May 1981).

B   Nuclear Energy and Weapons

Agarwal, A., ['Third World nuclear power'], *Nature*, vol. 279 (June 7, 1979), pp. 468–70.

Breach, I., *Windscale Fallout* (Harmondsworth, Middlesex: Penguin, 1978).

Connolly, T. J., *et al.*, *World Nuclear Energy Paths* (London: Royal Institute of International Affairs (RIIA), 1979).

Epstein, W., *The Last Chance* (New York: The Free Press, 1976).

Glasstone, S., *Sourcebook on Atomic Energy*, 3rd edn (Princeton: Van Nostrand, 1967).

Gottstein, K., 'Nuclear energy for the Third World', *Bulletin of the Atomic Scientist*, vol. 33 (1977), pp. 44ff.

Greenhalgh, G., *The Necessity for Nuclear Power* (London: Graham and Trotman, 1980).

Kaiser, K. (ed.), *Reconciling Energy Needs and Non-Proliferation* (Bonn: Europa Union Verlag GmbH, 1980).

Kaiser, K. and Lindemann, B. (eds), *Kernenergie und internationale Politik* (Munich: R. Oldenbourg, 1975).

Keeny, S. M., Jr., *et al.*, *Nuclear Power Issues and Choices* (Cambridge, Mass.: Ballinger, 1977).

Khan, M. A., *Nuclear Energy and International Cooperations:* [sic] *A Third World Perception of the Erosion of Confidence* (London: RIIA, 1979).

Lane, J. A., 'The impact of oil price increases on the market for nuclear power in developing countries', *IAEA Bulletin*, vol. 16 (April 1974), pp. 67–71.

Lawrence, R. M. and Larus, J., *Nuclear Proliferation: Phase II* (Lawrence: University Press of Kansas, 1974).

Lefever, E. W., *Nuclear Arms in the Third World* (Washington, D.C.: Brookings Institution, 1979).

Lönnroth, M. and Walker, W., *The Viability of the Civil Nuclear Industry* (London: RIIA, 1979).

Long, Hon. Clarence D., 'Nuclear proliferation: can Congress act in time?', *International Security*, vol. 2 (Spring 1977), pp. 52–76.

Mackerron, G., 'Third World energy', Letter to the editor, *New Scientist*, March 19, 1981, p. 765.

Mazrui, A. A., 'Changing the guards from Hindus to Muslims', *International Affairs*, vol. 57, no. 1 (Winter 1980/81), pp. 1–20.

Meyer-Wöbse, G., 'Nuclear cooperation in the Third World', *Aussenpolitik* (January 1978), pp. 65–74.

Park, J. K. (ed.), *Nuclear Proliferation in Developing Countries* (Seoul: Institute for Far Eastern Studies, 1979).

Pelayo, R. R., 'Financial aspects of nuclear power programs from the experience of Foratom member countries', mimeo. (Salzburg: IAEA, 1977).

Radetzki, M., *Uranium: A Strategic Source of Energy* (London: Croom Helm, 1981).

Sardar, Z., 'Why the Third World needs nuclear power', *New Scientist*, February 12, 1981, pp. 402–4.

Smart, I., Chairman, *Report of the International Consultative Group on Nuclear Energy* (London: RIIA, 1980).

Stockholm International Peace Research Institute, *World Armaments and Disarmament: SIPRI Yearbook, 1972–1979* (London: Taylor & Francis, 1973–80).

Williams, F. C. and Deese, D. A., *Nuclear Nonproliferation: The Spent Fuel Problem* (New York: Pergammon Press, 1979).

Woite, G., 'Can nuclear power be competitive in developing countries?', *Nuclear Engineering International*, July 1978, pp. 46–9.

Yager, J. A. (ed.), *Nonproliferation and U.S. Foreign Policy* (Washington, D.C.: Brookings Institution, 1980).

C   Argentina

Alemann, R., 'Economic development of Argentina', in Committee for Economic Development, *Economic Development Issues: Latin America* (New York: Praeger, 1967).

'Argentina's programme for '79 calls for major spending on energy infrastructure', *Business Latin America*, February 14, 1979, p. 51.

'CNEA's activities, plans', *La Prensa*, translated by Foreign Broadcast Information Service, December 29, 1978, p. 16.

Milenky, E. S., *Argentina's Foreign Policies* (Boulder, Co.: Westview Press, 1978).

'Nuclear power program in Argentina', *Nuclear Engineering International*, September 1971, p. 747.

'Offers, interest in Atucha-2 surveyed', *La Opinion*, translated by FBIS, December 29, 1978, p. 16.

D   Iran

'Atom-Abkommen Bonns mit Tehran, offenbar kurz vor dem Abschluss', *Frankfurter Allgemeine Zeitung*, February 9, 1976.

Cahn, A. H., 'Determinants of the nuclear option: the case of Iran', in O. Marwah and A. Schulz (eds), *Nuclear Proliferation and the Near-Nuclear Countries* (Cambridge, Mass.: Ballinger, 1975).

Clapp, G. R., 'Iran: a TVA for the Khuzestan region', *The Middle East Journal*, vol. 11 (Winter 1957), pp. 1–11.

Guery, C., 'Deux centrales nucléaires françaises pour l'Iran', *Figaro*, September 29, 1976, p. 1.

Guery, C., 'Feu vert pour deux centrales nucléaires françaises en Iran: un commande de 10 milliards de francs', *Figaro*, June 16, 1977.

Halliday, F., *Iran: Dictatorship and Development* (Harmondsworth, Middlesex: Penguin, 1979).

'Iran plans world's fourth biggest nuclear programme', *Nuclear Engineering International*, March 1977, pp. 31–2.

Lucas, N., 'Collapse of Iranian nuclear programme – big blow for Europe's nuclear industry', *European Energy Report*, February 1979, pp. 10–11.

E Indonesia

'Atomic energy: its future in Indonesia', *Indonesia Today*, December 1965, p. 11.

Hoadley, J. S., 'The politics of development planning agencies: the evolution of Indonesia's BAPPENAS', *Asia Quarterly*, no. 1 (1978), pp. 67–78.

Ichord, R. F., Jr., 'Indonesia', in G. J. Mangone (ed.), *Energy Policies of the World* (New York: Elsevier, 1977).

Jackson, K. D., 'Bureaucratic polity: a theoretical framework for the analysis of power and communications in Indonesia', in K. D. Jackson and L. Pye (eds), *Political Power and Communications in Indonesia* (Berkeley: University of California Press, 1978).

Saswinadi (ed.), *Supplement to the Proceedings: ITB [Institute of Technology at Bandung] – Industrial and Technical Research* (Bandung: BATAN, 1971).

Sethuraman, S. V., *Jakarta: Urban Development and Employment* (Geneva: International Labour Office, 1976).

Thomas, R. M., 'Indonesian science education and national development', in H. H. Beers (ed.), *Indonesia: Resources and Their Technological Development* (Lexington: The University Press of Kentucky, 1970).

F Latin America

'Brazil nuclear: shifting sands', *Latin America Weekly Report*, May 1, 1981, pp. 9–10.

Du Temple, O., 'Latin America: emerging nuclear market', *Nuclear News*, September 1979, pp. 59–64.

Gall, N., 'Atoms for Brazil, dangers for all', *Foreign Policy*, vol. 23 (Summer 1976), pp. 155–201.

Garcia Robles, A., *The Denuclearization of Latin America*, transl. by Marjorie Urquidi (Washington: Carnegie Endowment for International Peace, 1967).

Handler, B., 'Brazil plans 63 nuclear reactors this century', *Nature*, vol. 257 (October 9, 1975), p. 43.

Marchesi, I. H., 'Brazilian nuclear development program', mimeo., Brazilian Nuclear Energy Commission, April 1979.

'Mexico: green light for the energy program', *Latin America Weekly Report*, January 23, 1981, pp. 4–5.

Oteiza-Quirno, A., 'Technical assistance in Latin America', *IAEA Bulletin*, vol. 18, no. 3/4 (1976), pp. 24–8.

Redick, J. R., *Military Potential of Latin American Nuclear Energy Programs* (Beverly Hills: Sage Publications, 1972).

Redick, J. R., 'Regional restraint: US nuclear policy and Latin America', *Orbis*, vol. 22 (Spring 1978).

Rommel, W., 'The German Federal Republic in Latin America', *Studies in Developing Countries* (Warsaw), vol. 2 (1972), pp. 130–50.

G   *Asia*

Bray, F. T. J. and Moodie, M. L., 'Nuclear politics in India', *Survival*, May–June 1977, pp. 112–13.

Gonzaga, L., 'Back on track in the Philippines', *Far Eastern Economic Review*, October 17, 1980, pp. 62–3.

Jain, J. P., *Nuclear India* (New Delhi: Radiant, 1974).

Kanshik, B. M. and Mehrotra, O. N., *Pakistan's Nuclear Bomb* (New Delhi: Sopan, 1980).

Kapur, A., 'A nuclearizing Pakistan: some hypotheses', *Asian Survey*, vol. 20, no. 5 (May 1980), pp. 495–516.

Logarta, L. T., 'Hell all the way from Bataan and back', *Who* (Philippines), January 3, 1981, pp. 8–10.

Markettos, N. D., 'Update: ambitious nuclear programme planned for Taiwan', *Nuclear Engineering International*, March 1980, pp. 13–14.

Marwah, O., 'India's nuclear and space programs: intent and policy', *International Security*, vol. 1, no. 4 (Spring 1977), pp. 96–121.

Noorani, A. G., 'Indo-U.S. nuclear relations', *Asian Survey*, vol. 21, no. 4 (April 1981), pp. 399–416.

Pathak, K. K., *Nuclear Policy of India: A Third World Perspective* (New Delhi: Gitanjali Prakashan, 1980).

Poulouse, T. T. (ed.), *Perspectives of India's Nuclear Policy* (New Delhi: Young Asia Publications, 1978).

Subrahmanyam, K., *Nuclear Myths and Realities: India's Dilemma* (New Delhi: ABC Publishing House, 1981).

Subrahmanyam, K., *et al.*, 'Must India have the bomb?', *World Focus*, June 1981.

Tomar, R., 'The Indian nuclear power program: myths and mirages', *Asian Survey*, vol. 20, no. 5 (May 1980), pp. 517–31.

H   *Africa and Middle East*

Al-Gharem, H. E. A., 'Kuwait needs nuclear for desalination and power', *Nuclear Engineering International*, March 1977, pp. 35–7.

Eilam, U., 'The implementation of the nuclear power plant programme in Israel', *Nuclear Engineering International*, May 1977, pp. 59–62.

Heikal, M., *The Road to Ramadan* (London: Collins, 1975).

Jabber, F., *Israel and Nuclear Weapons* (London: Chatto & Windus, 1971).

Kurata, P., 'The outcasts [South Africa and Taiwan] forge new bonds', *Far Eastern Economic Review*, November 7, 1980, pp. 40–1.

Okoli, E. J., 'Nigeria going nuclear?', *West Africa*, November 10, 1980, pp. 2222–3.

Perera, J., 'Nuclear plants take root in the desert', *New Scientist*, August 23, 1979, pp. 577–80.

'Saddam Hussein and Arab nuclear technology', in *Iraq* (London: Iraqi Cultural Centre, June/July 1981), p. 1.

Wilmot, P. F., 'The future of Africa: revolution, retrogression or inertia', mimeo., Ahmadu Bello University [Nigeria] public lecture, April 23, 1981.

Index

Entries in italics refer to tables.